RECENT DEVELOPMENTS IN HIGH-ENERGY PHYSICS

Studies in the Natural Sciences

A Series from the Center for Theoretical Studies

University of Miami, Coral Gables, Florida

Orbis Scientiae: Behram Kursunoglu, *Chairman*

Recent Volumes in this Series

Volume 8 – PROGRESS IN LASERS AND LASER FUSION
Edited by Arnold Perlmutter, Susan M. Widmayer, Uri Bernstein, Joseph Hubbard, Christian Le Monnier de Gouville, Laurence Mittag, Donald Pettengill, George Soukup, and M. Y. Wang

Volume 9 – THEORIES AND EXPERIMENTS IN HIGH-ENERGY PHYSICS
Edited by Arnold Perlmutter, Susan M. Widmayer, Uri Bernstein, Joseph Hubbard, Christian Le Monnier de Gouville, Laurence Mittag, Donald Pettengill, George Soukup, and M. Y. Wang

Volume 10 – NEW PATHWAYS IN HIGH-ENERGY PHYSICS I
Magnetic Charge and Other Fundamental Approaches
Edited by Arnold Perlmutter

Volume 11 – NEW PATHWAYS IN HIGH-ENERGY PHYSICS II
New Particles–Theories and Experiments
Edited by Arnold Perlmutter

Volume 12 – DEEPER PATHWAYS IN HIGH-ENERGY PHYSICS
Edited by Arnold Perlmutter, Linda F. Scott, Mou-Shan Chen, Joseph Hubbard, Michel Mille, and Mario Rasetti

Volume 13 – THE SIGNIFICANCE OF NONLINEARITY IN THE NATURAL SCIENCES
Edited by Arnold Perlmutter, Linda F. Scott, Mou-Shan Chen, Joseph Hubbard, Michel Mille, and Mario Rasetti

Volume 14 – NEW FRONTIERS IN HIGH-ENERGY PHYSICS
Edited by Arnold Perlmutter, Linda F. Scott, Osman Kadiroglu, Jerży Nowakowski, and Frank Krausz

Volume 15 – ON THE PATH OF ALBERT EINSTEIN
Edited by Arnold Perlmutter and Linda F. Scott

Volume 16 – HIGH-ENERGY PHYSICS IN THE EINSTEIN CENTENNIAL YEAR
Edited by Arnold Perlmutter, Frank Krausz, and Linda F. Scott

Volume 17 – RECENT DEVELOPMENTS IN HIGH-ENERGY PHYSICS
Edited by Arnold Perlmutter and Linda F. Scott

ORBIS SCIENTIAE

RECENT DEVELOPMENTS IN HIGH-ENERGY PHYSICS

Chairman
Behram Kursunoglu

Editors
Arnold Perlmutter
Linda F. Scott

Center for Theoretical Studies
University of Miami
Coral Gables, Florida

PLENUM PRESS • NEW YORK AND LONDON

Library of Congress Cataloging in Publication Data

Orbis Scientiae, University of Miami, 1980.
Recent developments in high-energy physics.

(Recent developments in high-energy physics; v. 17)
Sponsored by the Center for Theoretical Studies, University of Miami, Coral Gables, Fla.
Includes index.
1. Particles (Nuclear physics)—Addresses, essays, lectures. I. Kursunoglu, Behram, 1922- II. Perlmutter, Arnold, 1928- III. Scott, Linda F. IV. Miami, University of, Coral Gables, Fla. Center for Theoretical Studies. V. Title. VI. Series.
QC793.28.072 1980 537.7'2 80-19774
ISBN 0-306-40565-2

Proceedings of Orbis Scientiae 1980
held by the Center for Theoretical Studies,
University of Miami, Coral Gables, Florida, January 14–17, 1980.

Printed in the United States of America

PREFACE

The editors are pleased to submit to the readers the state of the art in high energy physics as it appears at the beginning of 1980.

Great appreciation is extended to Mrs. Helga S. Billings and Mrs. Connie Wardy for their assistance with the conference and skillful typing of the proceedings which was done with great enthusiasm and dedication.

Orbis Scientiae 1980 received some support from the Department of Energy.

THE EDITORS

CONTENTS

The Variation of G and the Problem of the Moon............ 1
P.A.M. Dirac

Primatons, Maximum Energy Density Quanta, Possible Constituents of the Ylem........................... 9
A.J. Meyer

Dynamical Higgs Mechanism and Hyperhadron Spectroscopy..... 23
M.A.B. Bég

Glueballs: Their Spectra, Production and Decay............ 43
Sydney Meshkov

Quarks in Light Baryons................................... 61
Nathan Isgur and Gabriel Karl

Color van der Waals Forces?............................... 67
O.W. Greenberg

The Question of Proton Stability, Summary................. 87
Maurice Goldhaber

A Review of Charmed Particle Production in Hadronic Collisions................................ 89
Stephen L. Olsen

Infrared Properties of the Gluon Propagator: A Progress Report................................. 111
F. Zachariasen

Theoretical Aspects of Proton Decay....................... 121
William J. Marciano

Fun with E_6... 141
R. Slansky

SO_{10} as a Viable Unification Group........................ 165
P. Ramond

Migdalism Revisited: Calculating the Bound States of Quantum Chromodynamics........................... 173
Paul M. Fishbane

The U(1) Problem and Anomalous Ward Identities............ 189
Pran Nath and R. Arnowitt

Renormalizing the Strong-Coupling Expansion for Quantum Field Theory: Present Status........................ 211
Carl M. Bender, Fred Cooper, G.S. Guralnik, Ralph Roskies, and David Sharp

On the n-p Mass Difference in QCD.......................... 239
Geoffrey B. West

Testing Strong and Weak Gauge Theories in Quarkonium Decays.. 257
Boris Kayser

Selection Rules for Baryon Number Nonconservation in Gauge Models.. 277
R.E. Marshak and R.N. Mohapatra

Nonlinear Effects in Nuclear Matter and Self-Induced Transparency... 289
G.N. Fowler and R.M. Weiner

Program 299
Participants 303
Index 307

THE VARIATION OF G AND THE PROBLEM OF THE MOON

P.A.M. Dirac

Florida State University

Tallahassee, Florida 32306

What is the problem of the moon? It concerns the past history of the moon. The geological evidence that we have about it contradicts the astronomical evidence.

Rocks have been obtained from the surface of the moon and their ages have been determined by radioactive dating. The ages extend from 3.1 to 4.5 aeons. (One aeon = 10^9 years). There are none older than 4.5 aeons.

The oldest rocks found on earth are 3.9 aeons old. There is a chance that older ones may be discovered some day. This suggests that the earth-moon system was formed 4.5 aeons ago, and the oldest rocks on earth have been weathered away.

The meteorites all have the age 4.5 aeons. None more or less. So it looks as though the whole solar system was formed 4.5 aeons ago.

Now let us look at the astronomical evidence. The moon is circling the earth and its orbit is continually changing on account of tidal action. A high tide is produced on earth at approximately the point nearest to the moon and also at the antipodal point. Owing to the rotation of the earth, the axis of the tidal bulge is shifted around a little from the line joining the earth's center to the moon.

Let δ be the angle through which it is shifted. δ is a measure of the friction between the tidal bulge and the rotation of the earth below it. With no friction, δ would be zero.

The effect of the friction is to slow the rate of rotation of the earth. Angular momentum gets transferred from the earth to the orbital motion of the moon. The net result of the increase in the orbital angular momentum of the moon is that the moon's distance from the earth increases while its velocity decreases.

One can calculate the effect by using Newton's law of gravitation and conservation of angular momentum for the earth-moon system. One might also use Einstein's law of gravitation, but the difference would be negligible.

Various people have made calculations about the motion of the moon, using different degrees of approximation. I shall here use a calculation given by G.J.F. McDonald, which brings in the main features with neglect of the sun's influence. McDonald obtains the following formula for the distance a of the moon from the earth.

$$\frac{da}{dt} = \frac{3\, G^{1/2}\, m\, k_2\, R^5 \sin 2\delta}{a^{11/2}\, (M+m)^{1/2}} \quad . \tag{1}$$

Here M and m are the masses of the earth and moon, R is the radius of the earth, and k_2 is a Love coefficient of elasticity of the earth, which is a little less than 1.

The angle of shift δ is not a quantity that can be calculated with any accuracy, because it depends on the dissipation of tidal energy, which is not understood very well. It may change during the course of geological time. Its present value can be obtained from observations of the moon's motion. Using observations going back 250 years, one finds $\delta = 2.25^{\circ}$.

If we assume as a first approximation that δ is a constant, we can integrate equation (1). The result is

$$a = \left\{ \frac{39}{2} \frac{G^{1/2} m k_2 R^5 \sin 2\delta}{(M+m)^{1/2}} (t-t_0) + a_0^{13/2} \right\}^{2/13} , \tag{2}$$

with a_0 the present value of a and t_0 the present time.

Putting in numerical values, with t expressed in aeons, we get

$$\frac{39}{2} \frac{G^{1/2} m k_2 R^5 \sin 2\delta}{(M+m)^{1/2} a_0^{13/2}} = 0.6 \quad . \tag{3}$$

This leads to

$$a = \{ 0.6 (t-t_0) + 1 \}^{2/13} a_0 \; . \tag{4}$$

To apply this formula to a time 1.6 aeons ago we must put $t - t_0 = -1.6$. This gives a close to zero, so that the moon was very close to the earth. This happened at a time much more recent than the origin of the earth-moon system. If we go back still earlier, the formula obviously breaks down.

To understand what happened at this earlier time, McDonald makes a more elaborate calculation taking into account that the plane of the moon's orbit is inclined to the plane of the equator. The angle between them is now about 7°. The equations of motion show that this angle is decreasing. Thus it increases as one goes back into the past.

As one goes back to the time when the moon was very close to the earth, this angle increases to 90°. The moon then passed over the poles. At that time the moon's distance was about 3 earth radii.

If one goes back to still earlier times, the moon was in a retrograde orbit. Tidal action would then be making the moon approach the earth, instead of receding from it. So we get the picture of the moon originally in a retrograde orbit and slowly approaching the earth under the influence of tidal action. Then it drops down rapidly as the orbit becomes a polar one, and then the orbit becomes a direct one and tidal action causes the moon to recede, at first very rapidly and then slowly.

This picture of the moon's motion was deduced on the assumption that δ is constant. Maybe δ varied quite a bit, but any probable variation of δ would not change the character of the picture, but would just change the time of the close approach. Various people have worked on the problem using reasonable assumptions for the variation of δ, and they get results giving the time of close approach as somewhere between 1.5 and 2.5 aeons ago -- in any case much less than the age of the earth-moon system.

When the moon was very close to the earth, there must have been tremendous tides, not just a few meters high like at present, but several km high. This must have very much disturbed the formation of sedimentary rocks. There is not much evidence about ancient sedimentary rocks, but what there is shows no sign of such enormous tides, and geologists do not believe they ever existed.

This provides the problem of the moon. It could be that the moon was only formed recently, after the calculated time of close approach. But there is no plausible theory of its formation at this late date.

Many people have speculated about this problem without finding any plausible solution. J.G. Ford, in a review article published in 1975, brings out a new idea. He points out that the trouble may be ascribed to a lack of congruence between the time scales of the astronomers and geologists. The astronomers use a time scale, marked out by the motion of the earth around the sun, which they call ephemeris time. It is used with the equations of motion of Newton or Einstein. This might perhaps not be the same as the geological time scale, marked out by radioactive decay.

Let us suppose that, with respect to astronomer's time, radioactive decay in the past proceeded more rapidly than now. Then the rocks would appear older than they really are according to astronomer's time. The age of the moon, as given by radioactive dating, would be reduced and could perhaps be brought to less than the calculated time of close approach.

Ford put forward this idea, but did not like it very much. He says there is some geological support for it since sedimentation rates in the past (referred to radioactive clocks) appear to be less than now, but the figures he gives show a much greater effect than would follow just from the different time scales.

The time given by radioactive decay processes is presumably the same as atomic time, given by atomic clocks. This might be different from the ephemeris time used by astronomers. If there is a difference, one should be able to detect it by careful observations. One would have to observe some events in ephemeris time, observe them also with atomic clocks, and compare the results.

The best chance for making such observations is with the angular motion of the moon in the sky. This has been observed for some centuries with ephemeris time and has been observed since 1955 with atomic time. A comparison of these observations has been made by Van Flandern. Already by 1975 he reported a difference between the two times, but the difference had the wrong sign to help with the moon problem.

However, Van Flandern has been continually modifying his results, on account of improved calculations and more recent observational data. By 1978 the sign of the difference in the two times was reversed and it now helps with the moon problem.

If one is to follow up the idea of the two time scales, one needs a general theory to support it. Such a theory is provided by the Large Numbers hypothesis and the variation of G expressed in atomic units. I spoke about this theory at the conference last year.

With this theory we have to work with two metrics, an Einstein metric ds_E, which we must use with the equations of motion and an atomic and an atomic metric ds_A. They are related by

$$ds_E = (t/t_0)\, ds_A \,.$$

For studying the dynamics of the earth-moon system we must of course use the Einstein metric. Conservation of angular momentum will be valid with this metric.

We must now rewrite equation (1) to refer to the Einstein metric. It then reads

$$\frac{d\,a_E}{d\tau} = \frac{3\,G^{1/2}\,m\,k_2\,R_E^{\;5}\,\sin 2\delta}{a_E^{\;1/2}\,(M+m)^{1/2}} \quad . \tag{5}$$

Here the Einstein time τ, which is the same as ephemeris time, is connected with atomic time t by

$$\tau = \frac{1}{2}\,t^2/t_0 \quad .$$

The radius of the earth is determined mainly by atomic forces, so we may take R_A = constant. We then have

$$R_E = (t/t_0)R_A = (\tau/\tau_0)^{1/2}R_A = (\tau/\tau_0)^{1/2}\,R_0 \quad ;$$

where R_0 is the present radius. This brings a factor of $(\tau/\tau_0)^{5/2}$ into the right-hand side of equation (5), which changes the character of the solution drastically.

Instead of (2) we get

$$a_E = \left\{\frac{39}{14}\,\frac{G^{1/2}\,m\,k_2\,R_0^{\;5}\,\sin 2\delta}{(M+m)^{1/2}}\,t_0\left[\left(\frac{t}{t_0}\right)^7 - 1\right] + a_0^{\;13/2}\right\}^{2/13} \quad . \tag{6}$$

For t close to t_0 equation (6) goes over to equation (2).

Putting in the numerical values, we can use (3) again with R_0

instead of R and we find

$$a_E = \{0.5 \left[\left(\frac{t}{t_0}\right)^7 - 1 \right] + 1\}^{2/13} a_0 \quad . \tag{7}$$

The quantity in the { } brackets now never gets small, even if we go back to t = 0. The difference arises because of the factor R_E^5 in (5), with R_E::t, which makes much smaller tides in the past.

So the problem of the moon gets solved. The argument is not invalidated by quite considerable changes in the shift δ.

Thus the discussion of the past history of the moon provides evidence in favor of the variation of G and the Large Numbers hypothesis. The evidence is more positive than that provided by modern observations, which have to be carried out with extreme accuracy to show any effect at all.

REFERENCES

Dirac, P.A.M. On the Path of Albert Einstein p. 1 (1979), Orbis Scientiae, Center for Theoretical Studies, University of Miami, (Plenum Press, New York and London).

Ford, J.G. The Mercian Geologist Vol. 5 p. 205 (1975).

McDonald, G.J.F. The Earth-Moon System, Proceedings of Conference at the Goddard Space Flight Center, NASA (1964).

PRIMATONS

MAXIMUM ENERGY DENSITY QUANTA

POSSIBLE CONSTITUENTS OF THE YLEM

A.J. Meyer

The Chase Manhattan Bank, New York, N.Y. 10015

INTRODUCTION

The epistemological basis of quantum electrodynamics is weakened by the necessity of mathematically improper renormalizations associated with the existence of divergent integrals. These infinities have defeated attempts to apply standard mathematical principles to certain natural phenomena. This situation is not only annoying and mathematically absurd, but may have inhibited progress in many areas of theoretical physics. It would seem constructive to assume that these infinities are indeed perverse, that our model is incomplete and there may exist additional finite limits that have been overlooked.

Perhaps a fundamental step in this direction is to demonstrate that an invariant maximum density exists for electromagnetic radiation. Let us refer to this maximum energy density form as a primaton-photon. As I will show shortly, the primaton-photon may be defined as a photon whose three attributes: energy, spin, and "Compton" radius are sufficient to completely define an extreme Kerr black hole.

I'm not saying this primaton is a black hole, I'm just saying it uniquely maps or defines one.

From this beginning, a varied set of relationships may be developed. A few of the more interesting results of having a maximum density form for electromagnetic radiation are:

1. The primaton-photon may be likened to a noncollapsing black hole or symmetrically a nonexpanding white hole.
2. This maximum energy density suggests that the initial Gaussian curvature for the universe was finite and of order no greater than c^5/hG (i.e., no initial singularity).
3. Black hole formation can be represented by the collision of two primaton-photons, or two primaton-electrons. (Primaton-electrons are analogous to primaton-photons but are based on the more general Kerr Newman model.)
4. The primaton model may provide a more convenient way of looking at the relationships between the fundamental constants, e.g., G and the fine structure constant.
5. The primaton-photon may be representative of the heaviest virtual photons.

Today I will discuss the first point; provide a heuristic derivation of the maximum energy density conditions and discuss the formation of a black hole from black hole-like photons.

Several years ago, I began wondering if there did not exist common characteristics between black holes, photons and other elementary particles in limiting highly energetic states. Both uncharged black holes and photons may be completely characterized by their mass and spin. In the case of the photon the mass is its momentum divided by c, or effective mass, and the spin is one. For a photon, there exists a unique wavelength where the Compton radius is equal to the Kerr or gravitational radius of a maximally spinning black hole. This wavelength is of the order of 10^{-32} cm. It is exactly equal to the Planck length multiplied by 2π. In all energy states, the energy of the photon apparently resides at its Compton radius, which is equal to its effective radius of gyration. This situation suggests that photons are like naked ring singularities which become clothed in black at sufficiently high energies.

For other zero rest mass particles, such as neutrinos and gravitons, there is also a unique wavelength for which the outer Kerr radius is real and minimum (i.e., extreme Kerr hole radius). However, for these particles the energy does not reside at this radius. For the primaton form of the neutrino, the energy lies outside the radius of gyration, which is outside the extreme Kerr hole radius. This implies that neutrinos can never map into extreme Kerr holes. The converse is true for the graviton, when its energy is sufficient to map into an extreme Kerr hole - it becomes truly black hole-like. Table I illustrates this relationship for the triplet of primaton luxons. Notice that the graviton looks like a neutrino turned inside out. The trichotomy leads one to speculate as to whether neutrinos and gravitons are not the progeny of the primaton-photon. In fact, the name "primaton" was chosen since electromagnetic radiation in this form is a very likely candidate to be present in the primeval ylem before the "big bang" was 3.4×10^{-43} secs old.

However, the central focus of this presentation is to rough out a heuristic argument to show that one photon, the primaton-photon, possesses maximum energy density at a nonzero wavelength, and along the way attempt to point out some of its interesting properties.

THE PRIMATON-PHOTON

Consider the set of photons such that their wavelengths are equal to their effective Kerr circumferences. It is readily shown that this set of photons is unique, that is, they all have identical energies or wavelengths. It will be "shown" that the total energy of a primaton-photon resides at its extreme Kerr radius, implying that no collapse to a central ring singularity occurs.

Consider the class of photons for which the wavelength

Table I

Primaton particle	Extreme Kerr radius	Radius of gyration	Compton radius	Density
graviton	$\sqrt{2}R_\pi$	R_π	$R_\pi/\sqrt{2}$	$\rho_\pi/2$
photon	R_π	R_π	R_π	ρ_π
neutrino	$R_\pi/\sqrt{2}$	R_π	$\sqrt{2}R_\pi$	$\rho_\pi/4$

where R_π = Extreme Kerr radius of primaton-photon

ρ_π = Density of primaton-photon

$$\lambda = 2\pi r_s \ , \tag{1}$$

where

$$r_s = \frac{GM}{c^2} \pm \left[\left(\frac{Gm}{c^2}\right)^2 - \left(\frac{J}{mc}\right)^2 \right]^{\frac{1}{2}} , \quad \left(0 \leq |J| \leq \frac{Gm^2}{c}\right) \tag{2}$$

which gives the Kerr radii of the inner and outer event horizons of a rotating black hole as a function of its mass m, and angular momentum J. Where G is the universal gravitational constant and c = speed of light.

Now all photons have spin or angular momentum $J = \frac{h}{2\pi}$, where h = Planck's constant; (3)

and effective mass $m = \frac{h}{c\lambda}$. (4)

Combining equations (2), (3) and (4), we get

$$r_s = \frac{Gh}{c^3\lambda} \pm \left[\left(\frac{Gh}{c^3\lambda}\right)^2 - \left(\frac{\lambda}{2\pi}\right)^2 \right]^{\frac{1}{2}} . \tag{5}$$

By (1) and (5) the inner and outer event horizons are coincident and hence the extreme Kerr radius is

$$r_s = \left[\frac{Gh}{2\pi c^3}\right]^{\frac{1}{2}} = \text{Planck Length} \approx 1.616 \times 10^{-33} \text{cm.} \tag{6}$$

$R_\pi \equiv r_s$.

$$\lambda = \left[\frac{2\pi G}{c^3}\right]^{\frac{1}{2}} \quad \text{which is} \approx 1.01 \times 10^{-32} \text{cm} \tag{7}$$

by (4) and (7).

$$m = \left[\frac{hc}{2\pi G}\right]^{\frac{1}{2}} = \text{Planck mass} \approx 2.18 \times 10^{-5} \text{gm.} \tag{8}$$

Therefore, by (8) and (3),

$$J \equiv \frac{h}{2\pi} = \frac{Gm}{c^2}$$

which is the maximum angular momentum for a Kerr black hole.

Therefore, the primaton-photon, a photon with wavelength $\lambda = [\frac{2\pi G\hbar}{c^3}]^{\frac{1}{2}}$ is analogous to a maximally rotating Kerr black hole, which by (2) and (8) also has the minimum outer real event radius for any given mass.

Continuing on with this scalar mechanical treatment, an effective radius of gyration, r_g, for a photon is calculated as follows:

$r_g^2 = I/m$, where I is the photon's moment of inertia and m the effective mass. (9)

Now
$$J = I\omega = \frac{\hbar}{2\pi} \tag{10}$$

or
$$I = \frac{\hbar\lambda}{4\pi^2 c} \; . \tag{11}$$

By (9), (11) and (4) we get

$r_g = \frac{\lambda}{2\pi}$. In other words, for a photon of any energy, its effective radius of gyration is equal to its effective Compton radius. (12)

But by (1) and (12) when the photon is a primaton-photon

$r_g = r_s$; i.e., the extreme Kerr radius. (13)

This means the entire energy of the primaton-photon can also be thought of as residing at its Compton radius coincident with its extreme Kerr radius. Hence, no collapse to a central singularity occurs.

It has been suggested by Hawking that a white hole is equivalent to a time-reversed black hole. If indeed primatons are actually black holes, then since the primaton-photon is a luxon it has a null space-time separation. Hence, it would remain invariant under a time reversal. Therefore, certain primaton-luxons can also be thought of as white hole-like.

I admit that I'm taking liberties with the definition of a Kerr black hole by applying it to particles which have no inertial frame. But I'm using equations (1) and (2) in only a definitional sense, that is, photons with a wavelength determined by (1) and (2) are defined as primatons (black/white hole-like photons). Hopefully this definition has some empirical meaning as I will later try to illustrate in a "Gedanken Versuch".

PRIMATON-PHOTON DENSITY

In attempting to show that primaton-photons are the densest possible form of electromagnetic radiation, I am obviously not arguing that individual photons have invariant energy, but that the value for their maximum possible energy density is an invariant maximum, independent of inertial reference frame.

In other words there is no Lorentz transformation which can produce an energy density greater than that of a primaton-photon.

Let us define a photon's energy density as the ratio of the energy contained in its Compton sphere to its volume; when the Compton radius is less than the outer event radius we define the energy density in terms of the energy contained in the black hole with outer event radius, r_s.

That is, the energy density for a photon of wavelength λ can be represented by the continuous function $\rho(\lambda)$ where $\rho(\lambda)$ is finite everywhere over the interval $0 \leq \lambda \leq \infty$.

The derivative $\frac{d\rho}{d\lambda}$ is continuous everywhere except where $\rho(\lambda)$ is a maximum, which occurs

$$\text{at } \lambda = \left[\frac{2\pi G\hbar}{c^3}\right]^{\frac{1}{2}} \equiv \lambda_\pi \quad . \tag{14}$$

In other words, I believe a "reasonable" definition of a photon's energy density function $\rho(\lambda)$ can be given as follows:

$$\rho(\lambda) \equiv \begin{cases} \rho_c(\lambda) \quad \dfrac{\text{photon's energy}}{\text{Compton volume}} \; ; \text{ for all } \lambda \geq \lambda_\pi \\[2ex] \rho_s(\lambda) \quad \dfrac{\text{photon's energy}}{\text{volume defined by a rotating black hole having the same spin and energy as the photon}} \; ; \text{ for all } \lambda \text{ such that } 0 \leq \lambda \leq \lambda_\pi \; . \end{cases} \tag{15}$$

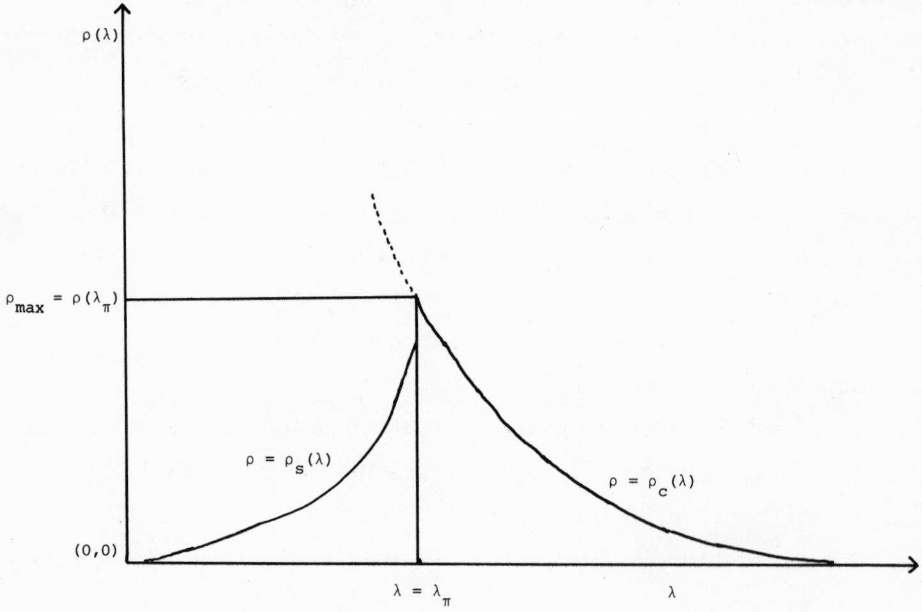

Figure 1

The graph of $\rho(\lambda)$ is shown in Figure 1. Using this definition of density we can demonstrate that a finite maximum exists.

It is easily shown that when $\lambda = \lambda_\pi$

$$\rho_c(\lambda_\pi) = \rho_s(\lambda_\pi) \ . \tag{16}$$

From (15) we have

$\rho_c(\lambda) = 3E(\lambda)/4\pi r_c^3$, where $E(\lambda) = \frac{hc}{\lambda}$ and $r_c = \frac{\lambda}{2\pi}$; i.e., the Compton radius. (17)

$\rho_s(\lambda) = 3E(\lambda)/4\pi r_s^3$ where $r_s(\lambda)$ is given by eq. (5) (18)

From (17) we get

$$\rho_c(\lambda) = \frac{6\pi^2 hc}{\lambda^4} \ , \ \lambda \geq \lambda_\pi \ . \tag{19}$$

From (14) we get

$$\rho_c(\lambda_\pi) = \frac{3c^7}{2hG^2} \ . \tag{20}$$

From (18) we get

$$\rho_s(\lambda) = \frac{3hc}{4\pi\lambda r_s^3(\lambda)}, \ 0 < \lambda < \lambda_\pi \ . \tag{21}$$

Now from (5) and (14) we get $r_s(\lambda_\pi) = [\frac{Gh}{2\pi c^3}]^{\frac{1}{2}}$. (22)

Therefore $\rho_s(\lambda_\pi) = \frac{3c^7}{2hG^2}$; (23)

thus the function $\rho(\lambda)$ is continuous at $\lambda = \lambda_\pi$ since

$$\rho_c(\lambda_\pi) = \rho_s(\lambda_\pi) = \frac{3c^7}{2hG^2} \ . \tag{24}$$

From (19) we can see that as λ increases, the energy density $\rho_c(\lambda)$ decreases inversely as the fourth power of λ. Therefore, $\rho_c(\lambda)$ is a maximum when λ is a minimum at $\lambda = \lambda_\pi$.

Expanding (21) we get

$$\rho_s(\lambda) = \frac{3hc}{4\pi\left(\frac{Gh}{c^3} + \left[\left(\frac{Gh}{c^3}\right)^2 - \frac{\lambda^4}{4\pi^2}\right]^{\frac{1}{2}}\right)\left(\frac{Gh}{c^3\lambda} + \left[\left(\frac{Gh}{c^3\lambda}\right)^2 - \left(\frac{\lambda}{2\pi}\right)^2\right]^{\frac{1}{2}}\right)^2} \tag{25}$$

for all λ such that $0 \leq \lambda \leq \lambda_\pi$.

By inspection of (25) one can see that the energy density $\rho_s(\lambda)$ will monotonically decrease from $\frac{3c^7}{2hG^2}$ to 0 as λ decreases from λ_π to 0. At $\lambda = \lambda_\pi$ the expression in the denominator of (25) has its minimum real value; for values of λ greater than λ_π, $\rho_s(\lambda)$ is complex. As the energy increases or equivalently as λ decreases the outer Kerr radius grows and the object approximates a nonrotating Schwarzschild black hole with Schwarzschild radius

$$r_s = \frac{2GE}{c^4} \quad . \tag{26}$$

From (25) we see that $\lim_{\lambda \to 0} \rho_s(\lambda) = 0$ so that when the object has infinite energy it has a density of zero. This is true for all black holes; if one similarly defines the density function in terms of the outer event radius.

BLACK HOLE-LIKE PHOTONS FORMING A SCHWARZSCHILD BLACK HOLE

The notion that certain photons are black hole-like probably has a ring of the bizarre about it. Therefore it might be helpful to do a "Gedanken Versuch" and illustrate the idea from different angle.

Let us consider a very high energy positron-electron collision resulting in two high energy photons having equal and opposite momentum vectors, but unable to escape from one another - in other words the collision forms a black hole consisting of two high energy photons. The rest mass of the system in the laboratory frame is given by

$$M_s = \frac{E_s}{c^2} = \frac{2h}{c\lambda_s} = \frac{2m_e}{\sqrt{1-\beta^2}}, \ \beta = \frac{v}{c}, \text{ where } m_e = \text{rest mass of electron.} \tag{27}$$

We now choose β such that when $t = \lambda/c$

$$ct = \lambda_s = R_s = \frac{2GM_s}{c^2}. \tag{28}$$

By (27) and (28)

$$\lambda_s = 2\left[\frac{Gh}{c^3}\right]^{\frac{1}{2}}. \tag{29}$$

From (14) and (29) we notice that the ratio of the primaton-photon wavelength to the wavelength of the photons comprising a black hole system is:

$$\frac{\lambda_\pi}{\lambda_s} = \frac{\sqrt{2\pi}}{2} > 1. \tag{30}$$

This means that the photons produced are truly black hole like since $\lambda_\pi > \lambda_s$. Indeed together they form a Schwarzschild black hole of total mass M_s and radius R_s.

$$M_s = \left[\frac{hc}{G}\right]^{\frac{1}{2}}; \quad R_s = 2\left[\frac{Gh}{c^3}\right]^{\frac{1}{2}}. \tag{31}$$

Each photon residing in the hole must have mass

$$m_s = \frac{M_s}{2} = \frac{1}{2}\left[\frac{hc}{G}\right]^{\frac{1}{2}}. \tag{32}$$

Now the mass of the primaton-photon is given by (8) as

$$m_\pi = \left[\frac{hc}{2\pi G}\right]^{\frac{1}{2}} \tag{33}$$

Table II

Primaton Π_i, i = 0,4	graviton Π_0	photon Π_1	neutrino Π_2	electron Π_3	proton Π_4
Charge q_i	0	0	0	$\pm e$	$\pm e$
Angular Momentum J_i	h/π	$h/2\pi$	$h/4\pi$	$h/4\pi$	$h/4\pi$
Rest mass m_{oi}	0	0	0	m_e	m_p
Primaton mass $M_{\pi i}$	$\sqrt{2}\ M_{\pi 1}$	$M_{\pi 1}$	$M_{\pi 1}/\sqrt{2}$	$\sqrt{\alpha_o}\ M_{\pi 1} \approx M_{\pi 2}$	$\sqrt{\alpha_o}\ M_{\pi 1} \approx M_{\pi 2}$
Extreme Kerr Radius R_{si}	$\sqrt{2}\ R_{\pi 1}$	$R_{\pi 1}$	$R_{\pi 1}/\sqrt{2}$	$\sqrt{\alpha_o}\ R_{\pi 1}$	$\sqrt{\alpha_o}\ R_{\pi 1}$
Radius of gyration R_{gi}	$R_{\pi 1}$	$R_{\pi 1}$	$R_{\pi 1}$	$R_{\pi 1}/\sqrt{\alpha_1}$	$R_{\pi 1}/\sqrt{\alpha_2}$
Effective Compton Radius = $\frac{\lambda_c}{2\pi} \equiv R_{ci}$	$R_{\pi 1}/\sqrt{2}$	$R_{\pi 1}$	$\sqrt{2}\ R_{\pi 1}$	$R_{\pi 1}/\sqrt{\alpha_o}$	$R_{\pi 1}/\sqrt{\alpha_o}$
Effective DeBroglie Radius = $\frac{\lambda_b}{2\pi} \equiv R_{bi}$	$R_{\pi 1}/\sqrt{2}$	$R_{\pi 1}$	$\sqrt{2}\ R_{\pi 1}$	$R_{\pi 1}/(\beta_3\sqrt{\alpha_o})$	$R_{\pi 1}/(\beta_4\sqrt{\alpha_o})$
v/c at extreme $\equiv \beta_i$	1	1	1	β_3	β_4
Maximum effective radius R_{mi}	$\sqrt{2}\ R_{\pi 1}$	$R_{\pi 1}$	$\sqrt{2}\ R_{\pi 1}$	$R_{\pi 1}/(\beta_3\sqrt{\alpha_o})$	$R_{\pi 1}/(\beta_4\sqrt{\alpha_o})$
Density at $R_{mi} \equiv \rho_{mi}$	$\rho_{\pi 1}/2$	$\rho_{\pi 1}$	$\rho_{\pi 1}/4$	$\beta_3^3\ \alpha_o^2\ \rho_{\pi 1}$	$\beta_4^3\ \alpha_o^2\ \rho_{\pi 1}$

Relationships and Definitions

(i) $M_{\pi 1} = [\frac{hc}{2\pi G}]^{\frac{1}{2}}$, $R_{\pi 1} = [\frac{Gh}{2\pi c^3}]^{\frac{1}{2}}$, $\rho_{\pi 1} = \frac{3c^5}{2hG^2}$ are the primaton-photons' mass, radius and density respectively.

(ii) $\alpha = \frac{2\pi e^2}{hc}$, fine structure constant

(iii) $\alpha_o \equiv [\alpha^2 + \frac{1}{4}]^{\frac{1}{2}}$

(iv) $\alpha_1 \equiv 2\alpha_o[1 - \alpha m_e^2 G/\alpha_o e^2]^{\frac{1}{2}}$; $\alpha_2 = 2\alpha_o[1 - \alpha m_p^2 G/\alpha_o e^2]^{\frac{1}{2}}$

(v) $m_e \equiv$ electron rest mass

(vi) $m_p \equiv$ proton rest mass $\approx (\frac{1}{\sqrt{2}})^{127} M_{\pi 1} = (\frac{1}{2})^{64} M_{\pi 0}$

(vii) $e = \sqrt{\alpha G}\, M_{\pi 1}$ = electron charge

(viii) $M_{eg} \equiv$ mass equivalent to charge, i.e., $GM_{eg}^2 = e^2$

(ix) $\sqrt{\alpha} = M_{eg}/M_{\pi 1}$, note that the ratio of strong to electromagnetic is $\approx \alpha^{-1} = [M_{\pi 1}/M_{eg}]^2$

(x) $M_{\pi 3} = M_{\pi 4} = \frac{e}{\sqrt{G}}[1 + (\frac{1}{2\alpha})^2]^{\frac{1}{2}} = M_{eg}[1 + (\frac{1}{2\alpha})^2]^{\frac{1}{4}} = \sqrt{\alpha}\, M_{\pi 1}[1 + (\frac{1}{2\alpha})^2]^{\frac{1}{4}} = \sqrt{\alpha_o}\, M_{\pi 1}$

(xi) $\beta_3 = [1 - m_e^2/M_{\pi 1}^2]^{\frac{1}{2}}$

(xii) $\beta_4 = [1 - m_p^2/M_{\pi 1}^2]^{\frac{1}{2}}$.

or

$$\frac{m_\pi}{m_s} = \frac{\sqrt{2\pi}}{\pi} < 1 \text{ or } m_s > m_\pi \ . \tag{34}$$

To build this Schwarzschild black hole required we need

$$\beta = \left(1 - \frac{4m_e^2 G}{hc}\right)^{\frac{1}{2}} \text{, which of course is almost unity.} \tag{35}$$

We should take note of the fact that Schwarzschild black holes "have no hair". There is no way to distinguish this hole from others with the same mass but formed differently. But in this case we had a positron-electron collision forming two black hole photons, which formed a stationary Schwarzschild black hole.

To conclude, Table II illustrates the primaton family. Where a primaton is defined as a particle π_i having attributes:

total energy	$M_{\pi i}c^2$
rest mass	m_{oi}
angular momentum	J_i
charge	q_i

such that

$$R_{si+} = \frac{GM_{\pi i}}{c^2} + \left[\left(\frac{GM_{\pi i}}{c^2}\right)^2 - \left[\left(\frac{J_i}{M_{\pi i}c}\right)^2 + \left(\frac{q_i^2}{M_{\pi i}c^2}\right)^2\right]\right]^{\frac{1}{2}} \tag{36}$$

is a minimum or equivalently, a primaton is a particle (M,J,q) such that

$$M^4 = (J^2c^2 + q^4)/G^2 \ . \tag{37}$$

DYNAMICAL HIGGS MECHANISM

AND HYPERHADRON SPECTROSCOPY*

M.A.B. Bég

The Rockefeller University

New York, New York 10021

ABSTRACT

A framework for implementing the Higgs mechanism -- in a dynamical way -- is described and discussed, and its experimental signatures are contrasted with those of the canonical theory with elementary spin-0 fields.

I. INTRODUCTION

It has been recognized for some time that the usual procedure for implementation of the Higgs mechanism, via introduction of elementary spin-0 fields in the Lagrangian, is one of the less satisfactory features of conventional Quantum Flavordynamics (QFD). The situation is acute in grand unified theories which seek to unify QFD and QCD in the framework of a single gauge theory. In such theories one must establish a hierarchy of nested gauge groups of the form

*Work supported in part by the U.S. Department of Energy under grant number EY-76-C-02-2232B.*000.

$$
\begin{aligned}
&U(1)_Q \otimes SU(3)_C \\
&\subset U(1) \otimes SU(2)_L \otimes SU(3)_C \\
&\subset - - - - - - - - \\
&\cdots\cdots\cdots\cdots\cdots\cdots \\
&\subset - - - - - - - - \\
&\subset \mathcal{G}
\end{aligned}
\qquad (1.1)
$$

where $U(1)_Q$ and $SU(3)_C$ are the gauge groups of QED and QCD respectively, $U(1) \otimes SU(2)_L$ is the Weinberg-Salam gauge group and $\mathcal{G}$ is the grand unification group. The groups indicated by dashes are the groups which, presumably, yield interesting physics in the energy interval 10^2 GeV - 10^{15} GeV; in other words, these are the groups which are likely to make the so-called desert[1] bloom. To establish this gauge-hierarchy, within the framework of the conventional methodology, one has to introduce multitudes of Higgs fields with judiciously contrived couplings; the essential simplicity of the gauge-theoretic approach is, thereby, irretrievably lost. The situation calls for a serious reconsideration of the suggestion[2-5] that one discard elementary Higgs fields altogether and seek to implement the Higgs mechanism in a dynamical way.

I shall discuss a possible scenario for dynamical symmetry-breaking and contrast its signature with that of the usual scheme involving elementary Higgs fields. Of particular interest will be the question of the experimental distinguishability of the two schemes, not only at the super-high energies at which there is an a priori expectation that differences will be manifest, but also at the relatively low energies required to produce Higgs-like spin-0 particles with masses ~10 GeV.[6] A noteworthy feature of the dynamical scenario is the emergence of a new hadronic spectroscopy in the TeV region.

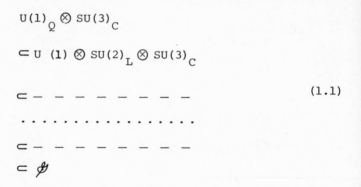

To put our discussion in the proper historical perspective, let us note that the effect customarily associated with the name of Higgs had in fact been discovered in 1958 by Anderson[7] in his study of the theory of superconductivity; indeed the work of Anderson provides the first, albeit nonrelativistic, example of a dynamical Higgs mechanism. Relativistic examples of this phenomenon (i.e. the spontaneous breakdown of local gauge invariance without the help of elementary spin-0 fields) are afforded by two-dimensional QED, solved by Schwinger[8] in 1962, and the U(1) models of Cornwall and Norton[9] and of Jackiw and Johnson.[10] Whether these examples illuminate the nature of dynamical symmetry breaking in the real world is open to question. The absence of a Goldstone theorem in two dimensions[11] renders suspect any general conclusions that one may be tempted to draw from two-dimensional models, the fact that symmetry-breaking in the U(1) models is an ultra-violet[12] effect makes it difficult to regard these models as paradigms relevant to the physical gauge hierarchy problem. At this time, the most promising approach to a dynamical alternative is based on the hope that QCD and QCD-like theories lead naturally to a (genuinely) spontaneous breakdown of chiral symmetries in flavor space; the resulting Goldstone bosons can then provide the longitudinal degrees of freedom that are needed to generate mass for flavor gauge fields. In this approach symmetry-breaking effects stem from the infra-red sector and one can meaningfully talk of short-distance or high-energy regimes in which a given symmetry may be deemed to have been restored.[13]

II. FLAVOR SYMMETRIES IN QCD

Some proven results and some plausible conjectures pertaining to flavor symmetries in QCD, which play a crucial role in our discussion, may be summarized as follows.

Let us consider the Lagrangian:

$$\mathcal{L}_{QCD} = -\frac{1}{8}\,\mathrm{Tr.}\ G_{\mu\nu}\ G^{\mu\nu}$$
$$+ \Sigma_f\ \overline{\psi}_f\ i\ \gamma^{\mu}\mathcal{D}_{\mu}\ \psi_f \qquad (2.1)$$
$$- \Sigma_f\ m_f\ \overline{\psi}_f\psi_f \quad ,$$

where G is the Yang-Mills curl of the suitably normalized color-gluon field matrix, $\mathcal{D}$ is the gauge-covariant derivative and the summation is over all N_F quark flavors.

In the limit in which all $m_f \to 0$, the accidental global invariance group of $\mathcal{L}_{QCD}$ in flavor space is

$$G_{QCD}\ (\mathrm{Acc.})$$
$$= SU(N_F)_L \otimes SU(N_F)_R \otimes U(1)_A \otimes U(1)_V \quad . \qquad (2.2)$$

The QCD vaccum is presumed to be such that the vector and axial-vector charges associated with the first two groups in Eq. (2.2) satisfy

$$\dot{Q}_i^V = 0 \quad , \qquad Q_i^V\ |0\rangle = 0 \quad , \qquad (2.3)$$

$$\dot{Q}_i^A = 0 \quad , \qquad Q_i^A\ |0\rangle \neq 0 \quad , \qquad (2.4)$$

where $i=1,2,\ldots,\ N_F^2-1$. Failure of the conserved axial charges to annihilate the vacuum implies the existence of N_F^2-1 Goldstone bosons and a dynamical or Nambu-Goldstone mass, M, for the quarks. That the mechanism leading to Eq. (2.4) is intrinsically nonperturbative is most easily seen by noting that M satisfies a homogeneous renormalization group equation; this equation can be solved in the limit of small coupling[14]:

$$M \propto \mu\ \exp\left(\frac{-c}{g^2}\right) \quad , \qquad (2.5)$$

where μ is the scale-parameter introduced to define the Green's functions of the theory, g is the renormalized QCD coupling constant and c is a known positive constant. It is possible, but has not been established to everyone's satisfaction, that Eq. (2.4) obtains in the presence of instantons[15]; at this time, it must be regarded as a plausible conjecture and its validity will be assumed in what follows. Note that f_π, the pion decay constant, also satisfies a homogeneous renormalization group equation; its g-dependence is therefore identical to that of M.

The generator of the $U(1)_A$ factor in Eq. (2.2) also fails to annihilate the vacuum; however the symmetry is broken explicitly by the 't Hooft mechanism[16] and the would-be flavor-singlet Goldstone boson becomes massive -- a process which will hereinafter be referred to as 't Hooftization[17]. Finally we note that the $U(1)_V$ factor in Eq. (2.2) merely corresponds to conservation of baryon number by the strong interactions and is not relevant to our considerations.

Now let us turn on the mass terms in the Lagrangian. The mass m_f is the mass that occurs in current algebra calculations and which may, in accordance with customary usage, be described as "current mass". [In QCD, as in QED, $m_f \bar{\psi}_f \psi_f$ is a finite operator; mass renormalization is therefore multiplicative and trivial and we shall not explicitly distinguish between bare and renormalized masses.] The net quark mass M_f, the so-called constituent mass, is a superposition of the current and the dynamical masses:

$$M_f = m_f + M + \ldots\ldots. \quad , \tag{2.6}$$

where the dots encompass terms such as the change in dynamical mass induced by the presence of current mass.

In any <u>specific gauge</u>, a precise definition of a momentum-dependent quark mass is furnished by

$$M_f(p) = \frac{1}{4} \mathrm{Tr}\, S'(p)_f^{-1} \quad , \tag{2.7}$$

where S' is the renormalized quark-propagator. The dynamical component of this mass is expected to damp rapidly (i.e. as a power of p) at high frequencies

$$M(p) \propto M. \left(\frac{p^2}{\mu^2}\right)^{-A} \tag{2.8}$$

whereas the current component varies only logarithmically

$$m_f(p) \propto m_f \cdot \{ \ell n \left(\frac{p^2}{\mu^2}\right)\}^{-B} \quad .$$

Here A is a dynamical parameter; B, however, is calculable in perturbation theory. Eqs. (2.8) and (2.9) permit us to refer to M and m_f as infra-red and ultra-violet masses respectively.

With the m_f's different from zero, the Goldstone -- identifiable with the ordinary π's, K's etc. -- pick up mass. Clearly, PCAC is a useful notion for flavors such that $m_f \ll M$. Since pion-PCAC is known to be good, we may assume that constituent u and d quark masses[18] provide us with a measure of the dynamical mass M:

$$M \simeq 3f_\pi \simeq 200 \text{ MeV} \quad . \tag{2.10}$$

Eq. (2.10) gives us the natural mass scale of QCD.

While the rest of the discussion will be predicated on the premise that the structure envisaged above -- for flavor symmetries in QCD -- carries over mutatis mutandis to other QCD-like theories, it should be stressed that our formulation can in no sense be regarded as complete; a major shortcoming is that it does not take account of the phenomenon of color-confinement.

III. THE HYPERCOLOR SCENARIO

The simplest strategy for dynamical generation of Goldstone bosons -- which can fulfill the needs of QFD and, hopefully, resolve the full gauge-hierarchy problem -- is to introduce a new species of quarks which come in at least two flavors with vanishing current-masses and are characterized by at least two, possibly four or more, new[19] colors C'. Strong interactions among the new quarks are generated by gauging $\mathcal{G}_{C'}$, the group associated with this (color)' degree of freedom; as indicated earlier, the structure of the resulting QC'D may be gleaned from analogy with QCD.

The (color)' degree of freedom, first introduced by Weinberg[4], has come to be known by a variety of names. Following ref. 6, we shall call it hypercolor; the terminology is convenient, with words such as hyperquark, hyperpion, hypersigma, etc. having an obvious meaning.

The weak and electromagnetic interactions of hyperquarks may be taken to be isomorphic to those of ordinary quarks, so that a flavor doublet such as (u',d') transforms under the electroweak gauge group in the same way as (u,d) or (c,s), etc.

With a single massless flavor doublet of hyperquarks, and $U(1) \otimes SU(2)_L$ as the electroweak group, the dynamical mechanism operates as follows. There are three massless hyperpions (the hyper-η is rendered massive by the 't Hooft mechanism); these mix with $W^{\pm}$ and Z which thereby acquire mass. The isotopic and hypercharge properties of the hyperpions, in conjunction with the isotopic-spin invariance of QC'D, imply:

$$m_Z^2 \cos^2 \xi = m_W^2 = \left(\frac{e}{2 \sin \xi} \right)^2 f_\pi^2 \quad , \tag{3.1}$$

where ξ is the Glashow-Weinberg-Salam angle ($\sin^2\xi \simeq 0.22$) and f_π, is the hyper-pion constant.

Consider, next, the case of several hyperquark doublets. While only the doublet (or doublets) with zero current mass can trigger the Higgs mechanism, i.e. move gauge-field masses from zero to finite values, doublets with nonzero current masses can enhance the masses of those gauge fields which have already acquired a finite mass. If we assume that all hyperpions -- irrespective of whether they belong to the same or different isotopic multiplets -- have equal decay constants and that they have masses which are either zero or negligible compared to W and Z masses, we may extend Eq. (3.1) as

$$m_Z \cos \xi = m_W = \frac{e}{2 \sin \xi} \cdot f_{\pi'} \sqrt{N_F'/2} \quad , \qquad (3.2)$$

where $N_F'/2$ is the total number of hyperquark doublets. It is tacitly assumed that the number of distinct flavors is exactly twice the number of electroweak doublets. Note that our assignment of hyperquarks to representations of the electroweak group is sufficient to ensure that the well-verified relationship between W and Z masses, a consequence of the use of Higgs doublets in the Weinberg-Salam model and often called the $\Delta I_{weak} = \frac{1}{2}$ rule, emerges naturally. Indeed, in so far as W and Z masses are concerned, the hypercolor scheme is equivalent to using an "elementary" Higgs doublet in the effective Lagrangian[4,6]

$$\mathcal{L}_{eff.} = (D_\mu \Phi)^\dagger \; (D^\mu \Phi) + - - - - , \qquad (3.3)$$

where D is the appropriate gauge-covariant derivative,

$$\Phi = \begin{pmatrix} i\pi'^{+} \\ \dfrac{\sigma' - i\pi'^{0}}{\sqrt{2}} \end{pmatrix} \qquad (3.4)$$

and

$$\langle \sigma' \rangle \equiv f_{\pi'} \sqrt{N_F'/2} \quad . \qquad (3.5)$$

Eqs. (3.3), (3.4) and (3.5) provide a concrete realization of the proposal that the Higgs fields of the canonical methodology be viewed as phenomenological props.[2] However, as we shall see later, the experimental signatures of elementary Higgses and the composite fields of the hypercolor scenario are <u>not</u> quite the same.[6]

The hyperpion decay constant, which determines the mass scale in QC'D, can be evaluated from Eq. (3.2):

$$f_{\pi'} \simeq \frac{250}{\sqrt{N_F'/2}}\ \mathrm{GeV} \quad . \tag{3.6}$$

The above-described dynamical scenario suffers from (at least!) two very serious difficulties if the overall gauge group is $G = U(1) \otimes SU(2)_L \otimes SU(3)_C \otimes \mathcal{G}_{C'}$: (A) In the absence of explicit Higgs fields, the accidental global invariance group of the Lagrangian is much larger than G. For example, with four quark flavors the Lagrangian is invariant under an additional $SU(2)_L$ group, the so-called horizontal group[20], as well as an $SU(4)_R$ group. Some of these extra flavor symmetries are broken, but only spontaneously, and therefore lead to unwanted Goldstone bosons. (B) No current quark masses or leptonic masses can ever be generated.

Attempts have been made to resolve the problems which afflict the hyper-color-based dynamical scheme; none can be said to have been terribly successful. To resolve (A), one may, for example, simply enlarge the flavor group until all unwanted Goldstone bosons can be gauged away; it is nontrivial, however, to make sure that the new flavordynamic interactions so generated are a negligible perturbation on the ordinary weak interactions at low energies. A satisfactory solution to problem (B) seems even more elusive. The simplest way to generate current quark masses is to enlarge the gauge group in such a way as to have currents which mix[21] C and C'; a current containing a piece of the form $(\bar{u}'\gamma_\mu u + \bar{d}'\gamma_\mu d)$, for example, can indeed lead to $m_u \neq 0$ (See Fig. 1). Similarly, for leptonic masses one needs C' -non-singlet currents which mix lepton number and baryon number. Current masses arising at the one loop level are expected to be of the order $\sim \alpha f_\pi$, perhaps a good starting point for understanding why these masses are small. However, in the simplest models, one either finds that $m_u = m_d$, $m_c = m_s$ etc. or one loses $m_W = m_Z \cos \xi$.

A challenging problem, then, is to find a grand unification group $\mathcal{G} \supset \mathcal{G}_F \otimes \mathcal{G}_C \otimes \mathcal{G}_C' \supset$ G such that: (a) $\mathcal{G}_F$ shrinks to its U(1) $\otimes$ $SU(2)_L$ sub-group in the low-energy limit. (b) All elementary fermions lie in an irreducible representation of $\mathcal{G}$ [This resolves difficulty (A)]. (c) All quarks and leptons are rendered massive in such a way that the $\Delta I_{weak} = \frac{1}{2}$ rule is not seriously jeopardized. While recognizing the possibility of bypassing the problem, at least in the form in which it is stated, I shall proceed on the assumption that a solution exists and that the mechanism which generates current fermion masses does not undercut the status of the group G, at the phenomenological level, in a quantitatively significant way. The very existence of a solution implies a rich particle spectrum coverning a substantial energy range, from a few GeV to a few TeV; to study this spectrum is our next logical task.

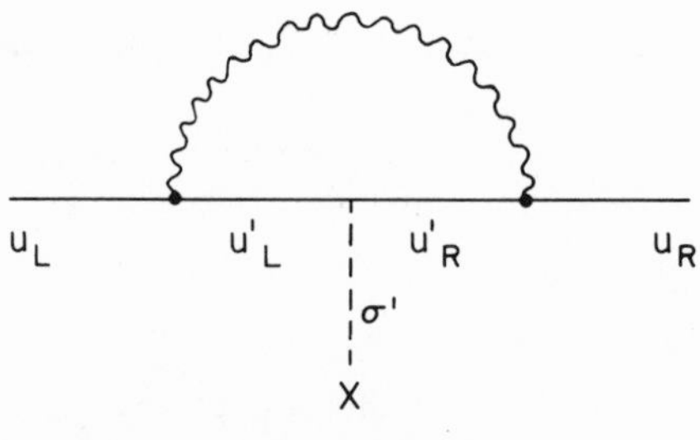

Fig. 1. Possible mechanism for generating current mass for the u-quark. The wavy line represents one of the heavy bosons which mediate sub-electroweak interactions.

IV. HYPERHADRON SPECTROSCOPY

To explore the spectroscopy of hyperhadrons, i.e. hypercolor-singlet hyper-quark, anti-hyperquark or multi-hyperquark states, it is necessary to know -- or, at least, have a lower bound on -- the number of hypercolors and have an educated guess for constituent hyperquark masses.

The number of hypercolors may be constrained by requiring that there exist a finite grand-unification mass $\mathcal{M}$ at which the QC'D and QCD coupling constants are of the same order:

$$\bar{g}_{C'}(\mathcal{M}) \approx \bar{g}_{C}(\mathcal{M}) \quad . \tag{4.1}$$

If $\mathcal{M} \gg \Lambda_C$ and Λ'_C, the characteristic mass scales in QCD and QC'D respectively, Eq. (4.1) is equivalent to

$$(11N'_C - 2N'_F)\, \ell n\left(\frac{\mathcal{M}}{\Lambda'_C}\right) \approx (11N_C - 2N_F)\, \ell n\left(\frac{\mathcal{M}}{\Lambda_C}\right) , \tag{4.2}$$

where N'_C and N'_F are, respectively, the number of colors and flavors in QC'D -- tacitly assumed to be an $SU(N'_C)$ gauge theory. In QCD, $N_C = e$ and I shall assume that $N_F = 6$. Since $\Lambda'_C \gg \Lambda_C$, it is evident that Eq. (4.2) can be satisfied only if

$$11N'_C - 2N'_F > 21 \quad . \tag{4.3}$$

Thus $N'_C \geqslant 3$ if $N'_F \leqslant 4$; if $N'_F \geqslant N_F$, $N'_C \geqslant 4$. Needless to add, sharper constraints would be obtained if one were to specify the grand-unification group $\mathcal{G}$ and the nature of $SU(N'_C)$ embedding in $\mathcal{G}$. In the following, where a specific value is needed, I shall take $N'_C = 4$.

To determine Λ'_C, which sets the scale of constituent hyper-quark masses, one may postulate that QC'D is a scaled-up version of

QCD, and equate the dimensionless ratios[6]

$$(f_{\pi'}/\Lambda'_C\sqrt{N'_C}) = (f_\pi/\Lambda_C\sqrt{N_C}) \quad . \tag{4.3}$$

Here $\sqrt{N_C}$ and $\sqrt{N'_C}$ are statistical color factors. The simplest argument for the occurrence of these factors, in the manner in which they do in Eq. (4.3), proceeds as follows. The charge-raising axial current associated with a prescribed flavor-doublet (u,d) is $A_\mu^{(-)} = \Sigma_\alpha\ \bar{u}_\alpha\ \gamma_5\gamma_\mu d_\alpha$ the summation being over all colors; the state vector of the pion with the same flavor quantum numbers is $\frac{1}{\sqrt{N_C}}\ \Sigma_\alpha\ |u_\alpha\ \bar{d}_\alpha>$. Hence, in a parton-like limit, $f_{\pi^+} \equiv f_\pi\sqrt{2}=f\ N_C/\sqrt{N_C}$, where f is a "reduced" Goldberger-Treiman constant for a given flavor and color. Since our scaling hypothesis is the statement that $(f'/\Lambda'_C) = (f/\Lambda_C)$, Eqs. (2.10), (3.6) and (4.3) imply

$$\Lambda'_C \approx 750\ \left(\frac{6}{N'_C\ N'_F}\right)^{1/2}\ \text{GeV} < 750\ \text{GeV} \quad , \tag{4.4}$$

the inequality following from the model-independent constraint on QC'D: $N'_C\ N'_F > 6$. To assess orders of magnitude, without getting involved in specific models, it is not unreasonable to set $\Lambda'_C \sim 0.5$ TeV.

Several facets of hyperhadron spectroscopy can now be discussed.

(i) Hyperbaryons: These are, by definition, multi-hyperquark states transforming according to the completely antisymmetric singlet representation of $SU(N'_C)$. The expected masses are, therefore, in the neighborhood of

$$M_B \sim N'_C\ \Lambda'_C \approx 1.8\ \left(\frac{N'_C}{N'_F}\right)^{1/2}\ \text{TeV} \quad . \tag{4.5}$$

Conservation of hyperbaryon number is expected to be on the same footing as conservation of ordinary baryon number; consequently, hyperbaryons have to be pair-produced. Experimental searches for these objects must therefore await the construction of machines yielding $\sim$ 4 TeV in the center-of-mass frame. In the meantime, hyperbaryons offer a fertile field for theoretical conjecture and speculation, and some interesting problems for cosmologists. Perhaps the most important question stems from the possibility that hypernucleons (defined to be the lowest mass hyperbaryons) and hypernuclei may constitute new islands-of-stability in the hadronic spectrum in the trans-TeV region. A substantial amount of such matter should have been created in the early stages of evolution and one may legitimately ask: where is this hypermatter?

(ii) Hyper-Vector-Mesons: If the current masses of quarks and hyperquarks are of the same order of magnitude -- a feature found in many proposed models for mass-generation -- almost all hyperhadronic mass is dynamical. This means that the wide disparity between ρ-ω, J/ψ and Υ masses is unlikely to be duplicated in the hypermeson spectrum; one expects, rather, that all vector meson masses would be close to

$$M_{(VM)'} \cong 2\Lambda'_C \quad . \tag{4.6}$$

The hyper vector mesons will undoubtedly decay fairly rapidly into relatively light pseudoscalars (see below); however, a credible estimate of the widths is not available at this time. An e^+e^- machine with 0.5-1 TeV beams (Super LEP?) would afford the best means for producing and studying these 1^- hypermesons.

(iii) Hyper-Scalar-Mesons: The set of 0^+ hypermesons (mass ~1 TeV) includes the σ', the precise analogue of the left-over Higgs of the Weinberg-Salam model. However -- unlike the Higgs particle -- the σ' will decay very rapidly into light pseudoscalars and, in consequence, will be much too broad to be visible as a distinct particle.

(iv) Hyper-Pseudoscalar-Mesons and The Pseudo-Goldstone Sector: Pseudoscalar mesons are, by far, the most important particles in the hyper-hadron spectrum; their importance stems from the fact that they are the lightest hyperparticles -- the only ones that can be produced at energies which, if not currently available, are likely to be available in the near future.

In the absence of any mechanism for generating current hyper-quark masses which are anywhere more than a small fraction of the dynamical masses, the pseudoscalar mesons are best viewed in terms of the Nambu-Goldstone phenomenon -- rather than as $^1S_0$ states of a naive hyperquark model. The modes generated by the QC'D Lagrangian, in the presence of electroweak and other (sub-electro-weak) interactions, may be categorized as follows: (a) True Goldstone bosons; these -- unless they are electrically neutral and have other special properties such that they cannot be ruled out on the basis of existing experiments -- should not occur. (b) True (would be) Goldstone bosons, absorbed in the Higgs mechanism (c) Goldstone bosons subject to 't Hooftization. (d) Pseudo-Goldstone[22] bosons which acquire mass through the agency of electro-weak and sub-electroweak interactions. It is this last category which concerns us here.[23]

The masses of the hyper pseudo-Goldstone bosons, hereinafter generically labelled π', may be estimated in two different ways. The simplest way is to note that since the mass arises at the one loop level:

$$m^2_{\pi'} \sim \alpha \, f^2_{\pi'} \simeq \left(\frac{20}{\sqrt{N'_F/2}} \text{ GeV} \right)^2 \quad , \tag{4.7}$$

so that

$$m_{\pi'} \lesssim 20 \text{ GeV} \quad . \tag{4.8}$$

A somewhat less naive estimate is obtained by comparison of near-Goldstone modes in QC'D and QCD, using current algebra supplemented with our N_C-scaling hypothesis:

$$m_{\pi'}^2 = m_\pi^2 \; \frac{m_{q'}}{m_q} \cdot \frac{f_{\pi'}}{f_\pi} \cdot \left(\frac{3}{N_C'} \right)^{1/2} , \tag{4.9}$$

or

$$m_{\pi'} = 7 \text{ GeV} \; \left(\frac{m_{q'}}{m_q} \right)^{1/2} \left(\frac{6}{N_C' \, N_F'} \right)^{1/4} \tag{4.10}$$

$$\lesssim 7 \text{ GeV for } m_{q'} \sim m_q \quad . \tag{4.11}$$

The electromagnetic radius of charged hyperpions may be inferred from Eq. (4.6); it is $\sim\sqrt{6}\,(2\Lambda_C')^{-1} \sim (0.4\text{ TeV})^{-1}$. Consequently, at energies of a few tens of GeV the hyperpions are essentially point-like and should be produced copiously in e^+e^- annihilation; their contribution to the usual R-parameter is

$$\Delta R = \frac{1}{4} \Sigma_{\pi'} \, Q_{\pi'}^2 \; \left(1 - \frac{4m_{\pi'}^2}{s} \right)^{3/2} , \tag{4.12}$$

where the summation extends over all hyperpions with charges $Q_{\pi'}$, which can be produced at c.m. energy $\sqrt{s}$. The best way to detect hyperpions would be to study events of the form $e^+e^- \to \tau^\pm$ + (hadrons) + $\nu \to \mu^\pm$ + (hadrons) + ν's.

V. EXPERIMENTAL DISTINGUISHABILITY OF DYNAMICAL AND ELEMENTARY-HIGGS SCENARIOS AT LOW ENERGIES

While experiments in the trans-TeV region -- when they become possible in the not-too-near future -- would shed much light on the nature of the Higgs mechanism, it is important to know whether experimental signals in the 10-100 GeV range can be used to distinguish to two scenarios.

Let us note first that the hyperpion mass-range, Eqs. (4.8)

and (4.11), overlaps with that for the left-over Higgs of the Weinberg-Linde bound[24] for the latter mass is $m_\phi > 7$ GeV. It seems possible, therefore, that the pseudo-Goldstone modes of the dynamical theory can mask differences between the two schemes by mocking the Higgses of the conventional methodology. Closer scrutiny shows, however, that this need not be the case.

To appraise the distinctive features of hyperpions and elementary Higgses, let us compare the effective interactions of the former with the couplings of the latter:

$$\mathcal{L}_{\text{eff.}} \left\{ \overline{\begin{pmatrix} a \\ b \end{pmatrix}} \begin{pmatrix} a \\ b \end{pmatrix} \right\}$$

$$= G_F f_{\pi'} (-i)\bar{a} \; [(m_a - m_b) - \gamma_5 (m_a + m_b)] \, b\pi'^{+} + \text{H.C.} \tag{5.1}$$

$$+ G_F f_{\pi'} \sqrt{2} \, [m_a \bar{a} i \gamma_5 a - m_b \bar{b} i \gamma_5 b] \, \pi'^{0} \quad ,$$

$$\mathcal{L}_1 = \sqrt{G_F \sqrt{2}} \; [\, m_a \bar{a} a + m_b \bar{b} b \,] \, \phi^0 \quad , \tag{5.2}$$

$$\mathcal{L}_2 = \tilde{\mathcal{L}}_1 + \frac{1}{\sqrt{2}} \quad (f_1 \bar{a} a + f_2 \bar{b} b) \, \eta^0$$

$$+ (- f_1 \bar{a} i\gamma_5 a + f_2 \bar{b} i \gamma_5 b) \, \chi^0 \,]$$

$$+ \frac{1}{2} \bar{a} \; [(f_2 - f_1) + \gamma_5 (f_2 + f_1)] \; b \, \chi^{+} + \text{H.C.} \quad . \tag{5.3}$$

Here a and b are quarks or leptons such that $\begin{pmatrix} a \\ b \end{pmatrix}_L$ is an electroweak doublet, $\mathcal{L}_1$ corresponds to the Weinberg-Salam model with one complex doublet (one physical spin-0 field, ϕ^0) and $\mathcal{L}_2$ corresponds to the same model with two complex doublets (five physical spin-0 fields: ϕ^o, η^o, χ^o and $\chi^\pm$). Also, G_F is the Fermi constant and f_1 and f_2 are unknown parameters.

Consider first the problem of distinguishing π'^o from ϕ^o. Both have universal couplings -- in the sense that a trivial mass rescaling enables one to go from one electroweak doublet to another -- of comparable semi-weak strength:

$$G_F\, f_{\pi'}\, \sqrt{2} \,/\, \sqrt{G_F\,\sqrt{2}} \simeq \sqrt{2/N_{F'}} \quad .$$

However, to leading order in the weak interactions, $\pi'^o \to f\,\bar{f}$ is an S-wave decay whereas $\phi^o \to f\,\bar{f}$ is P-wave. The decay rates and branching ratios are, therefore, different. Furthermore, if we look at exclusive hadronic channels, $\pi'^o \to D\,\bar{D}\,\pi$ and $\phi^o \to D\,\bar{D}$ are allowed whereas $\phi^o \to D\,\bar{D}\,\pi$ and $\pi'^o \to D\,\bar{D}$ are forbidden. Yet another handle (to be discussed more fully elsewhere) is afforted by processes such as π'^o or $\phi^o \to 2$ color gluons $\to 2$ hadronic jets; measurements on the jets can provide information about gluon polarization and thence about the initial state. Two photon decays can also yield information about parity; however, the 2γ branching ratios appear to be disappointingly small.

Next, let us consider the problem of distinguishing hyperpions from χ-like particles. No qualitative selection-rule criteria are available here; however, one may attempt to resolve the problem by exploiting the following differences: (a) π'-couplings exhibit universality whereas (in the absence of additional, unnatural or artificial, input) χ-couplings do not. (b) The χ-fields are necessarily accompanied by relatively light 0^+ neutrals (ϕ^o, η^o) whereas the π'-fields are not so encumbered.

The extent of our ability to ascertain, through experiments in the 10-100 GeV range, whether nature prefers the hypercolor-based dynamical scenario or the canonical formuation in terms of elementary Higgs fields may be encompassed in the following statements:

(i) The presence of ϕ^o-like particles, with scalar couplings at semi-weak strength, would be decisive evidence against the QC'D-based dynamical scheme.

(ii) If the couplings of spin-0 particles to fermions exhibit universality (modulo trivial mass factors) and if the light neutrals are all pseudoscalars, the dynamical picture may be deemed to be preferred.

VI. CONCLUDING REMARKS

The hypercolor-based dynamical scheme holds considerable promise; when intergrated into a well-defined procedure for doing precise calculations, it may provide a natural resolution of the gauge-hierarchy problem in grand unified theories -- and liberate us from the rather awkward burden of the Higgs sector. While many profound problems (some of which have been underlined above) are blocking progress at this time, it is remarkable that some experimental implications of the dynamical mechanism can be stated before the theoretical problems have found a satisfactory resolution. Indeed, it is not inconceivable that the nature of the Higgs mechanism will be pinpointed by our experimental colleagues in the not-too-distant future. And if nature signals a preference for dynamical symmetry-breaking, we would have a strong motivation for proceeding to the TeV region to explore a whole new spectroscopy.

ACKNOWLEDGMENTS

It is a pleasure to thank L. Dolan, M. Gell-Mann, S. Meshkov, H.D. Politzer and P. Ramond for enjoyable discussions on various aspects of dynamical symmetry breaking.

REFERENCES AND FOOTNOTES

1. S.L. Glashow, (Private communication).
2. M.A.B. Bég and A. Sirlin, Annu. Rev. of Nuc. Sci. 24, 379 (1974). [See, especially, the last paragraph of Section 7.2].

3. M.A.B. Bég, in New Frontiers in High Energy Physics, Proceedings of Orbis Scientiae 1978, edited by A. Perlmutter and L. Scott (Plenum, New York, 1978).
4. S. Weinberg, Phys. Rev. D13, 974 (1976).
5. L. Susskind, SLCA Report No. 2142, 1978 (Phys. Rev. D, to be published).
6. M.A.B. Bég, H.D. Politzer and P. Ramond, Phys. Rev. Lett. 43, 1701 (1979).
7. P.W. Anderson, Phys, Rev. 112, 1900 (1958).
8. J. Schwinger, Phys. Rev. 128, 2425 (1962).
9. J.M. Cornwall and R.E. Norton, Phys. Rev. D8, 3338 (1973).
10. R. Jackiw and K. Johnson, Phys. Rev. D8, 2386 (1973).
11. S. Coleman, Comm. in Math. Phys. 31 259 (1973).
12. Cf. L. Dolan and R. Jackiw, Phys. Rev. D9, 3320 (1974).
13. M.A. B. Bég and S. -S. Shei, Phys. Rev. D12, 3092 (1975).
14. See, for example, D. J. Gross and A. Neveu, Phys. Rev. D10, 3235 (1974).
15. C.G. Callan Jr., R. Dashen and D.J. Gross, Phys. Rev. D17, 2717 (1978); D. Caldi, Phys. Rev. Lett. 39, 121 (1977); R.D. Carlitz, Phys. Rev. D17, 3225 (1978).
16. G. 't Hooft, Phys. Rev. Lett. 37, 8 (1976).
17. The apt terminology is due to M. Gell-Mann, (Private communication).
18. Constituent u and d masses ~300 MeV may be inferred from quark-model calculations of the absolute values of proton and neutron magnetic moments [M.A.B. Beg, B.W. Lee and A. Pais, Phys. Rev. Lett. 13, 514 (1964)] if the quarks are presumed to have no anomalous or Pauli magnetic moment. For two very different procedures for assessing the mass scale in QCD see: M.A.B. Bég, Phys. Rev. D11, 1165 (1975) and H. Georgi and H.D. Politzer, ibid. D14, 1829 (1976).
19. The possibility of getting by without new colors, by introducing exotic quarks belonging to higher (6,8,10 ---- etc.) represen-

tations of $SU(3)_C$ has been discussed by W. Marciano, Rockefeller University Report No. COO-2232B-193 (1980).

20. This horizontal group was labelled $SU(2)_K$ in: M.A.B. Bég, Phys. Rev. D8, 664 (1973). [This paper appears to have been the first to discuss the possibility that interactions generated by gauging this group, the so-called "horizontal interactions", may be of physical relevance. For other discussions of such interactions, see: M.A.B. Bég and A. Sirlin, Phys. Rev. Lett. 38, 1113 (1977); M.A.B. Bég and H.-S. Tsao, ibid. 41, 279 (1978); S. Barr and A. Zee, Phys. Rev. D17, 1854 (1978); F. Wilczek and A. Zee, Phys. Rev. Lett. 42, 421 (1979)].
21. S. Dimopoulos and L. Susskind, Nucl. Phys. B155, 237 (1979).
22. S. Weinberg, Phys. Rev. Lett. 29, 1698 (1972).
23. Apart from a few minor changes, such as keeping the number of hyperquark flavors arbitrary, the discussion of hyper-pseudo-Goldstone bosons follows that in ref. 6.
24. S. Weinberg, Phys. Rev. Lett. 36, 294 (1976); A.D. Linde, JETP Letters 23, 73 (1976).

GLUEBALLS: THEIR SPECTRA, PRODUCTION AND DECAY

Sydney Meshkov

National Bureau of Standards

Washington, D.C. 20234

What is a glueball? It is a colorless and flavorless composite of two or more gluons whose existence is implied in quantum chromodynamics (QCD) [1-3]. That we are even considering such an object today is a true tribute to serendipity. Let us see why I say that.

Nineteen years ago, Gell-Mann and Ne'eman suggested using SU(3) to describe the interactions and spectroscopy of baryons and mesons (figure 1). In principle, there could have been states with exotic quantum numbers in addition to the 1's, 8's and 10's of baryons, and the 1's and 8's of mesons. The lack of exotics led to the introduction of quarks by Gell-Mann and Zweig. Once quarks were proposed, it was clear that they could have spin. This leads to SU(6). All of the observed mesons and baryons were described as composites of $q\bar{q}$,L and qqq,L.

The lowest baryon states, 56,L = 0 seemed to have the wrong statistics. Systems of fermions, such as quarks, should be totally antisymmetric under the interchange of all of the coordinates of the particles. Nevertheless, the 56 is symmetric. To rectify this, a bugger factor, now called color was introduced by Wally Greenberg. Each quark was given an additional SU(3) degree of freedom, color, such that when we compounded the color

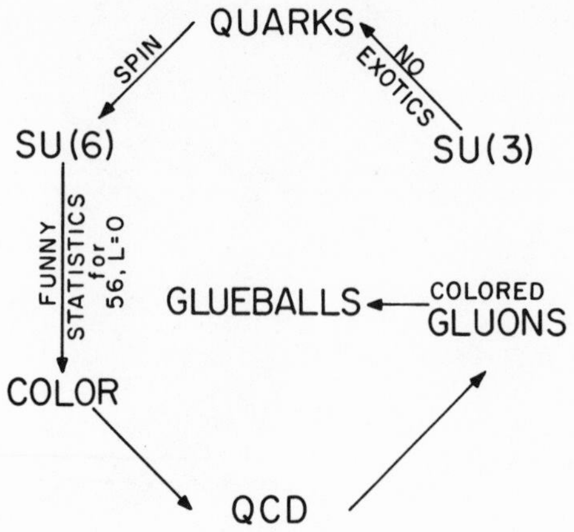

Figure 1. The Serendipitous Path to Glueballs.

indices of 3 quarks we got an antisymmetric 1. Remember: $3 \times 3 \times 3 = 10_{\text{sym.}} + 8 + 8 + 1_{\text{antisymmetric}}$. Now the Pauli Principle is satisfied. Physical states are color singlets.

Once we had the concept of color, a non-abelian gauge theory, QCD, analogous to QED, was constructed. In QCD, massless colored gluons (color octets) shuttle between colored quarks and antiquarks as carriers of the strong force.

Since the gluons are color 8's and since 8×8 couple to a symmetric 1 and $8 \times 8 \times 8$ couple to two 1's, one symmetric 1_D and one antisymmetric 1_F it looks like there should be color singlet nonquark states which are composites of gluons. We call these states Glueballs.

The work on glueballs that I am discussing today has been done in collaboration with J. Joseph Coyne of the National Bureau of Standards and with Paul M. Fishbane of the University of Virginia. In this work we first make a systematic enumeration of the quantum numbers of the two gluon and three gluon states. New states not allowed in the usual quark model of mesons are, in the two gluon sector $J^{PC} = 1^{-+}, 3^{-+}, \ldots$; and in the three gluon sector $J^{PC} = 0^{+-}, 2^{+-}, \ldots$; and 0^{--}. Next we discuss various production and decay modes of these unique states. In addition, in a more dynamical vein, mixing of glueballs with ordinary quark states has important consequences. For example, such mixing gives a radiative decay width of the ψ/J into a 0^{-+} glueball comparable to its radiative decay width into the η_c. This result is surprisingly independent of the location of the glueball relative to the η_c. We then estimate the magnitude of this decay using the OZI rule and suggest that a search for such decays would be a sensitive test for the existence of glueballs. The location of glueball states is a crucial question which we do not address here. There are many predictions which cover a wide range of masses, but we are unable to choose between them. Finally, we discuss the properties of hyperballs, composites of hypergluons of the color prime group SU(4).

Strictly speaking, there is no gauge invariant separation of glueballs into 2-gluon, 3-gluon, etc., sectors, because the Lagrangian of QCD permits gluon self-couplings. This would also be true of the multiquark (exotic) sector of the theory. A gauge covariant description of glueballs is possible using the $F_{\mu\nu}$ tensor of QCD, which is equivalent in the ground state to superposition of color electric fields E(1^{--} states) and color magnetic fields B(1^{+-} states). The states of this type corresponding to no spatial excitation have been listed by Bjorken [3]. Excitations of such states may be generated by the action of gauge covariant derivatives on the lowest-lying states, or by considering products of $F_{\mu\nu}$'s. Recent work indicates [4] that the latter procedure may be correct. Nevertheless, as is usual in QCD, we find it more convenient as well as physically appealing to consider a break-up of our states according to the number of gluons.

Let me review the way that we construct the two gluon states, for pedagogical purposes, even though it has all been said by Robson and others [1-3]. The principal requirement that we impose is that the two gluon wave functions be totally symmetric under the interchange of the space, spin and color coordinates of the gluons. We take each gluon to be a $J^{PC} = 1^{--}$ object. The two gluon system is $1^{--} \times 1^{--}$ in a relative L wave. The color part of the wave function is always symmetric, since $8 \times 8 \to 1_D$. This tells us that only C = + states are allowed. Therefore only two sets of states are allowed: either states which are symmetric in spin (S = 0,2) and symmetric in space (L even, $P = (-)^L$) or states which are antisymmetric in spin (S = 1) and antisymmetric in space (L odd).

Since $J = L + S$, the allowed states are:

$$
\begin{aligned}
L &= 0{:}\ 0^{++},\ 2^{++} \\
L &= 1{:}\ 0^{-+},\ 1^{-+},\ 2^{-+} \\
L &= 2{:}\ 2^{++},\ 0^{++}\ \ldots\ 4^{++} \\
L &= 3{:}\ 2^{-+},\ 3^{-+},\ 4^{-+} \qquad .
\end{aligned}
$$

A comparison with the usual $\bar{q}q$,L quark model (table I) shows that for quarks there are no C = odd states and odd^{-+} states. Two gluon states have no C = odd states, but the odd^{-+} sequence 1^{-+}, 3^{-+}, ... is allowed. As we shall see below, all J^{PC} states are allowed for three gluons.

The full enumeration of the three-gluon system necessarily involves the assumption of an effective potential, as in [5-7] the three-quark model of baryons. Our use of a shell model is motivated by its success in describing quark systems. We require total symmetry in the three-gluon color-singlet wave function, which consists of spatial excitations, a spin part, and a color part. For the latter, a color singlet is possible with either a purely symmetric (S) d-type coupling or a purely anti-symmetric (A) f-type coupling; mixed symmetry (M) [8] is not allowed. The charge

Table 1

List of states (J^{PC}) for quark model and glueballs. √(×) indicates allowed (disallowed). 2g (3g) correspond to glueball states formed from 2 (3) gluons.

State	$q\bar{q}$	2g	3g
0^{++}	√	√	√
0^{+-}	x	x	√
0^{-+}	√	√	√
0^{--}	x	x	√
1^{++}	√	√	√
1^{+-}	√	x	√
1^{-+}	x	√	√
1^{--}	√	x	√
2^{++}	√	√	√
2^{+-}	x	x	√
2^{-+}	√	√	√
2^{--}	√	x	√
3^{++}	√	√	√
3^{+-}	√	x	√
3^{-+}	x	√	√
3^{--}	√	x	√

conjugation properties of the system depend only on the symmetry of the color wave function; d coupling gives C = - and f coupling gives C = +.

The spin wave functions for products of three 1^{--} gluons can in general have S, A, or M symmetry. The pure S states may have spins S = 1 or 3; the M states may have S = 1 or 2; and the pure A state has S = 0.

The spatial wave functions are labelled $(n_1\ell_1)(n_2\ell_2)(n_3\ell_3)$, where n is a principal quantum number and ℓ is an orbital angular momentum. These spatial wave functions also have a definite symmetry character and associated total orbital angular momenta L which combine with the total spin S to give definite J. Because of the intrinsic negative parity of each gluon, the total parity of each state is $(-1)^{\ell_1+\ell_2+\ell_3+1}$.

Table 2 summarizes the results for the three-gluon system up to the $(1s)^2(1d)$ configuration. Depending on the central potential, however, many of the 105 states shown may be spurious, corresponding to motion of the overall center of mass of the system. For example, for the harmonic oscillator [6], 51 of these states are spurious. This is partly due to the degeneracy of the $(1s)(1p)^2$, $(1s)^2(2s)$ and $(1s)^2(1d)$ levels in the harmonic oscillator. All states are nonspurious for $(1s)^3$. For $(1s)^2(1p)$, only those states with M spatial symmetry are allowed. For the $(1s)(1p)^2$, $(1s)^2(2s)$, $(1s)^2(1d)$ complex, allowed states arise from five linear combinations and all others are spurios. The spatial symmetries, L values, nonspurious combinations of configurations, and their contents are:

(a) symmetric, $L = 0, \sqrt{\frac{2}{3}}\,(1s)^2(2s) + \sqrt{\frac{1}{3}}\,(1s)(1p)^2$: 1^{--}; 3^{--}; 0^{-+}

(b) mixed, $L = 0, \sqrt{\frac{1}{3}}\,(1s)^2(2s) + \sqrt{\frac{2}{3}}\,(1s)(1p)^2$: 1^{--}; 2^{--}; 1^{-+}; 2^{-+}

(c) symmetric, $L = 2,\ \sqrt{\frac{2}{3}}\,(1s)^2(1d) - \sqrt{\frac{1}{3}}\,(1s)(1p)^2$: 2^{-+}; $1,2,3^{--}$; $1,\ldots 5^{--}$

(d) mixed, $L = 2,\ \sqrt{\frac{1}{3}}\,(1s)^2(1d) - \sqrt{\frac{2}{3}}\,(1s)(1p)^2$: $1,2,3^{--}$; $0,\ldots 4^{--}$; $1,2,3^{-+}$; $0,\ldots 4^{-+}$

(e) antisymmetric, $L = 1$, $(1s)(1p)^2$: 1^{--}; $0,1,2^{-+}$; $2,3,4^{-+}$.

The four combinations orthogonal to (a) - (d) are spurious, as is the mixed $L = 1$ $(1s)(1p)^2$ state. No states with unique sets of quantum numbers are eliminated by this procedure. In general, the determination of which states are spurios depends on the particular central potential in which the gluons are moving and may be very difficult to determine.

The following remarks are of relevance.

(a) How might we distinguish a multi-gluon glueball from the corresponding multi-$(\bar{q}q)$ exotic? Exotics can have isospin, charge and flavor, while glueballs cannot [9].

(b) We may form $(q\bar{q})g$ color singlets by coupling the $(q\bar{q})$ in a color octet. For no spatial excitation between the g and the $(q\bar{q})$ system, we can form every J^{PC} combination except 0^{+-}. With one unit of orbital angular momentum this gap can be filled.

(c) In the four gluon sector for color SU(3), there are no A color singlets, only S and M color singlets.

Listed below are some possibly attainable production and decay modes for $J^{PC} = 0^{+-}$, 0^{--}, and 1^{-+} states, all of which have no $(q\bar{q})$ counterpart. Wherever possible we have tried to use the unique signatures provided by photons. First, for production,

Table 2

Shell model of glueballs made of 3 gluons. The allowed states for composites of 3 gluons, each in a potential, are given for various configurations. The allowed states have wave functions that are totally symmetric under the interchange of space, spin and color degrees of freedom.

con-figur.	spat. sym.	orbit. ang.mom.	spin sym.	spin	color sym.	J^{PC}
$(1s)^3$	*S*	0	*S*	1,3	d	$1^{--};3^{--}$
	S	0	*A*	0	f	0^{-+}
$(1s)^2(1p)$	*S*	1	*S*	1,3	d	$0,1,2^{+-};2,3,4^{+-}$
	S	1	*A*	0	f	1^{++}
	M	1	*M*	1,2	d	$0,1,2^{+-};1,2,3^{+-}$
	M	1	*M*	1,2	f	$0,1,2^{++};1,2,3^{++}$
$(1s)(1p)^2$	*S*	0,2	*S*	1,3	d	$1^{--};3^{--};1,2,3^{--};1...5^{--}$
	S	0,2	*A*	0	f	$0^{-+};2^{-+}$
	M	0,2	*M*	1,2	d	$1^{--};2^{--};1,2,3^{--};0,...4^{--}$
	M	0,2	*M*	1,2	f	$1^{-+};2^{-+};1,2,3^{-+};0,...4^{-+}$
	M	1	*M*	1.2	d	$0,1,2^{--};1,2,3^{--}$
	M	1	*M*	1,2	f	$0,1,2^{-+};1,2,3^{-+}$
	A	1	*S*	1,3	f	$0,1,2^{-+};2,3,4^{-+}$
	A	1	*A*	0	d	1^{--}
$(1s)^2(2s)$	*S*	0	*S*	1,3	d	$1^{--};3^{--}$
	S	0	*A*	0	f	0^{-+}
	M	0	*M*	1,2	d	$1^{--};2^{--}$
	M	0	*M*	1,2	f	$1^{-+};2^{-+}$
$(1s)^2(1d)$	*S*	2	*S*	1,3	d	$1,2,3^{--};1,...5^{--}$
	S	2	*A*	0	f	2^{-+}
	M	2	*M*	1,2	d	$1,2,3^{--};0,...4^{--}$
	M	2	*M*	1,2	f	$1,2,3^{-+};0,...4^{-+}$

$$1^{++}(\chi_1) \to \gamma + 0^{+-} \qquad \text{(P wave)}$$

$$1^{++}(\chi_1) \to \gamma + 0^{--} \qquad \text{(S wave)}$$

$$0^{-+}(\eta_c) \to \gamma + 0^{--} \qquad \text{(P wave)}$$

$$1^{++}(\chi_1) \to 1^{-+} + 1^{-+} \qquad \text{(S wave)} \quad .$$

For decay, we list

$$0^{+-} \to \gamma + 1^{++}(D) \qquad \text{(P wave)}$$

$$0^{--} \to \gamma + 1^{--}(\omega) + 1^{--}(\omega) \qquad \text{(S wave)}$$

$$\to 3\omega$$

$$1^{-+} \to \gamma + 1^{--}(\omega) \qquad \text{(P wave)}$$

$$\to 0^{-+}(\eta) + 0^{-+}(\eta') \qquad \text{(P wave) [2]} \quad .$$

Even though we have emphasized the existence of glueball states which have no counterpart in the quark model, it nevertheless seems that a useful way of demonstrating the existence of glueballs is to detect effects due to the mixing of quark model states with their glueball counterparts. We construct a simple model* of the mixing of two 0^{-+} states, one the η_c and one a pseudoscalar glueball, and show that as a result we expect a significant decay of the ψ/J into photon plus glueball even though the masses are insensitive to this (small) mixing.

Let us begin with our two states, labeled with C and G corresponding to the η_c 0^{-+}, largely made from a $c\bar{c}$ quark pair, and the corresponding 0^{-+} glueball. We write any mass as $m = \bar{m} - i\frac{\Gamma}{2}$, where $\bar{m}$ is the real part and $\frac{\Gamma}{2}$ the imaginary part (Γ = width) of

*K. Ishikawa (reference 1) was the first to recognize this possibility.

of the mass. We use a superscript ^{o} to indicate the unmixed state, and no superscript for the physical (mixed) state.

The mixing is introduced through a coupling constant f, dimensions of $(\text{mass})^2$, which connects a C with a G propagator. Then the full propagator for, say, a C state of momentum q would have the form

$$\langle C|\Delta|C\rangle = \frac{1}{q^2-m_C^{o2}} + \frac{1}{q^2-m_C^{o2}}\, f\, \frac{1}{q^2-m_G^{o2}}\, f\, \frac{1}{q^2-m_C^{o2}} + \ldots$$

$$= \frac{1}{q^2-m_C^{o2}}\left(1 - \frac{f^2}{(q^2-m_G^{o2})(q^2-m_C^{o2})}\right)^{-1} . \qquad (1)$$

The 2 × 2 matrix for the propagator of the full C, G system is then

$$\Delta = \left(1 - \frac{f^2}{(q_G^2-m_G^{02})(q_C^2-m_C^{o2})}\right)^{-1} \begin{bmatrix} (q^2-m_C^{o2})^{-1} & f(q^2-m_C^{o2})^{-1}(q^2-m_G^{o2})^{-1} \\ f(q^2-m_C^{02})^{-1}(q^2-m_G^{o2})^{-1} & (q^2-m_G^{o2})^{-1} \end{bmatrix} \qquad (2).$$

Inverting

$$\Delta^{-1} = \begin{bmatrix} q^2-m_C^{o2} & -f \\ -f & (q^2-m_G^{o2}) \end{bmatrix} . \qquad (3)$$

To find the physical states, we diagonalize the mass matrix

$$M^2 = q^2 - \Delta^{-1} = \begin{bmatrix} m_C^{o2} & f \\ f & m_G^{o2} \end{bmatrix} . \qquad (4)$$

The elements of the diagonalized M^2 (the eigenvalues λ_1, λ_2 of eq. (4)) are the physical masses m_C^2 and m_G^2. The eigenvectors determine the physical states in terms of the unmixed states.

Since the decay of the ψ/J into γ plus G can occur through an admixture of C^o in G, we are not interested in the full eigenvectors but only in the mixing angle

$$\tan\theta \approx \frac{f}{\bar{m}_C^{o2} - \bar{m}_G^{o2}} \quad . \tag{5}$$

We find the mixed masses by an approximation method: the eigenvalues λ_1 and λ_2 are written

$$\lambda_1 = m_C^{o2} + 2m_C^o\,\delta_1 \quad , \quad \lambda_2 = m_G^{o2} + 2m_G^o\,\delta_2 \quad ,$$

where $\delta_1 << m_C^o$, $\delta_2 << m_G^o$. This gives

$$\delta_1 \approx \frac{f^2}{2m_C^o(m_C^{o2} - m_G^{o2})} \quad , \quad \delta_2 \approx -\frac{f^2}{2m_G^o(m_C^{o2} - m_G^{o2})} \quad . \tag{6}$$

Separating real and imaginary parts,

$$\begin{aligned}
\bar{m}_C^2 &\approx \bar{m}_C^{o2} + \frac{f^2}{\bar{m}_C^{o2} - \bar{m}_G^{o2}} \\
\Gamma_C &\approx \frac{\bar{m}_C^o}{\bar{m}_C}\left(\Gamma_C^o - \frac{f^2(\bar{m}_C^o\Gamma_C^o - \bar{m}_G^o\Gamma_G^o)}{\bar{m}_C^o(\bar{m}_C^{o2} - \bar{m}_G^{o2})^2}\right) \\
\bar{m}_G^2 &\approx \bar{m}_G^{o2} - \frac{f^2}{\bar{m}_C^{o2} - \bar{m}_G^{o2}} \\
\Gamma_G &\approx \frac{\bar{m}_G^o}{\bar{m}_G}\left(\Gamma_G^o - \frac{f^2(\bar{m}_C^o\Gamma_C^o - \bar{m}_G^o\Gamma_G^o)}{\bar{m}_G^o(\bar{m}_C^{o2} - \bar{m}_G^{o2})}\right) \quad .
\end{aligned} \tag{7}$$

Note that the splitting of the physical states is, of course, increased over that of the bare states.

Equations (5) and (7) could be used in a calculation of the branching ratio of the ψ/J into $\gamma + G(0^{-+})$. We avoid much of the dynamics by instead calculating only the ratio

$$R \equiv \frac{\Gamma(\psi/J \to \gamma + G(0^{-+}))}{\Gamma(\psi/J \to \gamma + \eta_c)} .$$

This ratio consists of three parts: First, a Gaussian overlap factor describing the wave function at the origin as a function of the photon momentum k. This factor is of the form [10] $\exp(- k^2 \frac{r^2}{3})$, where r is a scale which we take to be $r = 2\ \mathrm{GeV}^{-1}$. Second, since the decay is P-wave, we have a factor $(k_G/k_C)^3$. Finally we have the mixing angle $\tan\theta$. In summary,

$$R = \left(\frac{k_G}{k_C}\right)^3 \exp\left(- \frac{r^2}{3} (k_G^2 - k_C^2)\right) \frac{f}{\bar{m}_C^{o2} - \bar{m}_G^{o2}} , \qquad (8)$$

where

$$k_G = \frac{m_{\psi/J}^2 - m_G^2}{2m_{\psi/J}} , \qquad (9)$$

and similarly for k_C.

The masses are determined by f^2, whereas the mixing angle, and hence R, samples f. This means we may expect very small mass and width shifts with, nevertheless, a substantial value of R. Such sampling of configuration mixing by probing the wave function rather than mass shifts is well-known in both atomic and nuclear physics.

For a numerical test, we take [11] $\bar{m}_C = 2.98$ GeV, $\Gamma_C = 0.020$ GeV, and vary $\bar{m}_G^o$, Γ_G^o, and f. As a result, $\bar{m}_C^o$, Γ_C^o, $\bar{m}_G$, and Γ_G, as well as $\tan\theta$, are then determined. We run $\bar{m}_G^o$ from 1.2 to 2.8 GeV. Γ_G^o runs from 20 to 100 MeV, but its variation has very little effect on R. The value of $\Gamma_G \approx \Gamma_G^o$ is important only in the sense that for a given R, the decay is harder to detect experimentally for a broad G than for a narrow one.

Table 3 shows R for various values of $\bar{m}_G^o$ and for various f^2. R is remarkably independent of $\bar{m}_G^o$ given its three components. Over the entire range of values, and indeed much further, neither the position nor the width of either the G or the C states varies by as much as 0.1%. Moreover, R is independent of the total width of the G state, as eq. (8) shows.

Let us now ask what value for f is reasonable. We may estimate the magnitude of f^2 by considering the M1 transition $\psi \to \gamma + \eta_c$, compared to $\psi \to \gamma + \eta'$ or $\psi \to \gamma + \eta$. Since there is an f^2 factor in the η or η' decay width to lowest order, but not in the η_c decay width, we expect, for example, $\Gamma(\psi \to \gamma + \eta_c)/\Gamma(\psi \to \gamma + \eta) = \mathcal{O}((f/m_\eta)^{-2})$ times the phase space and overlap factors of eq. (8). Experimental values required here are [12] $BR(\psi \to \gamma + \eta) = (1.17 \pm 0.17) \times 10^{-3}$, $BR(\psi \to \gamma + \eta') = (6.87 \pm 1.71) \times 10^{-3}$, and with less certainty [13] $BR(\psi \to \gamma + \eta_c) \cong 0.01$. This argument gives values for f which run from 0.05 - 0.25 GeV^2. Another estimate for f comes from purely hadronic processes. A typical OZI-violating decay, where one quark flavor pair converts to another (i.e., proportional to f^2) has a rate of about 1% of an OZI-allowed process.* Thus, f^2 on an hadronic scale of 1 GeV^2 is approximately 0.01 or $f \cong 0.1\ GeV^2$. These two estimates for f thus seem reasonably consistent. For $f^2 \lesssim 0.01$, table 3 shows that R will be four or less, without much variation as a function of m_G for a given f. (The fact that R is so large is an effect of the combination of phase space, overlap, and $\tan\theta$.) Such values of R could be missed if the G-state is sufficiently

*This argument also suggests that the hadronic decay widths of a glueball is the geometric mean of an ordinary OZI-allowed hadronic decay and the OZI-violating decay of, say, charmonium into ordinary hadronic matter, i.e., between 10 and 30 MeV.

Table 3

Values of $R = \frac{\Gamma(\psi \rightarrow \gamma + G)}{\Gamma(\psi \rightarrow \gamma + \eta_c)}$ for various values of M_G, the glueball mass, and various strengths of coupling, f^2.

M_G \ f^2	.005	.010	.015	.020
1.2	1.52	2.16	2.64	3.05
1.3	1.65	2.34	2.86	3.30
1.4	1.79	2.53	3.10	3.58
1.5	1.93	2.73	3.35	3.86
1.6	2.08	2.94	3.60	4.16
1.7	2.22	3.14	3.85	4.44
1.8	2.35	3.33	4.08	4.71
1.9	2.46	3.49	4.27	4.93
2.0	2.54	3.60	4.41	5.09
2.1	2.57	3.64	4.46	5.15
2.2	2.54	3.60	4.41	5.10
2.3	2.45	3.46	4.24	4.91
2.4	2.27	3.21	3.94	4.56
2.5	2.02	2.86	3.50	4.06
2.6	1.70	2.41	2.96	3.42
2.7	1.33	1.90	2.33	2.71
2.8	.97	1.39	1.73	2.01

broad, say 100 MeV or more. On the other hand, R is sufficiently large to make a careful search both feasible and desirable. There would seem to be ample room in the total radiative decay width of the ψ/J to accommodate the decay $\psi/J \rightarrow \gamma + G$.

The above is one example of a possible series of calculations of this type. Others might include mixing of a gluonic 1^{--} state with the ψ/J itself which would be manifested in the decay

$$^3P_1(1^{++}) \rightarrow \gamma + G(1^{--}) \qquad \text{(S wave)}$$

$$^3P_2(2^{++}) \rightarrow \gamma + G(1^{--}) \qquad \text{(S wave)} \quad .$$

In a recent paper [14] on dynamical symmetry breaking the Goldstone bosons, which furnish the longitudinal degrees of freedom for massive gauge fields are considered to be bound states of a new species of quark, called hyperquarks. These hyperquarks which have superstrong interactions (generated by gauging a color' degree of freedom) spontaneously break chiral symmetry. The color' degree of freedom is called "hypercolor;" composites of the hyperquarks, hyperpions and hypersigmas may be formed. If hypercolor corresponds to the SU(4) of color' [15], then we may form hyperballs, in a manner completely analogous to our treatment of glueballs. For SU(4), as for SU(3), only two color singlets, 1_D and 1_F may be formed from 3 gluons. In fact, only 1_D and 1_F singlets may be formed for 3 gluons for any SU(n) of color, $n > 3$. This means that the spectroscopy for hyperballs is the same as for glueballs, except for an overall mass scale shift. According to reference [14], this scale may be in the TeV range.

ACKNOWLEDGEMENTS

We are indebted to the Aspen Center for Physics, where this work started, for its hospitality. We thank M.A.B. Beg, J.D. Bjorken, E.D. Bloom, F. Bulos, C.E. Carlson, J.M. Cornwall,

H.J. Lipkin, J. Noble, A. Polyakov, P. Ramond, J. Rosner, J. Schwarz, Gordon Shaw, and A. Soni for their help, suggestions and very stimulating discussions.

REFERENCES

[1] H. Fritzch and P. Minkowski, Nuovo Cimento 30A (1975) 393. R.L. Jaffe and K. Johnson, Phys. Lett. 60B (1976) 201. P.G.O. Freund and Y. Nambu, Phys. Rev. Lett. 34 (1975) 1645. J. Kogut, D.K. Sinclair and L. Susskind, Nucl. Phys. B114 (1976) 199. P. Roy and T.F. Walsh, Phys. Lett. 78B (1978) 62. K. Koller and T. Walsh, Nucl. Phys. B140 (1978) 449. K. Ishikawa, Phys. Rev. D20 (1979) 731. J.F. Bolzan, W.F. Palmer, and S.S. Pinsky, Phys. Rev. D14 (1976) 3202. H. Suura, Univ. of Minn. Preprint, submitted to Phys. Rev. Lett.

[2] D. Robson, Nucl. Phys. B130 (1977) 328.

[3] J.D. Bjorken, SLAC-PUB-2366, (August 1979). J.D. Bjorken, SLAC Summer Institute on Particle Physics (1979).

[4] P.M. Fishbane, S. Gasiorowicz, and P. Kaus, in preparation.

[5] O.W. Greenberg, Phys. Rev. Lett. 13 (1964) 598. R.H. Dalitz, in Les Houches Lectures, 1965 (Gordon and Breach), New York 1965).

[6] D. Faiman and A.W. Hendry, Phys. Rev. 173 (1968) 1720.

[7] N. Isgur and G. Karl, Phys. Lett. 72B (1977) 109, N. Isgur and G. Karl, Phys. Lett. 74B (1978) 353.

[8] M. Hamermesh, Group Theory (Addison-Wesley, Reading, Mass., 1962).

[9] Gordon Shaw, private communication.

[10] J. Rosner, private communication.

[11] T. Burnett, Crystal Ball Results, Irvine Conference, Dec. 1979.

[12] R. Partridige, et al., SLAC-PUB-2430 (November 1979), (submitted to Physical Review Letters).

[13] C. Peck, Invited talk, Annual Meeting of the American Physical Society, McGill University, Montreal, Canada, (October 25-27, 1979).

[14] M.A.B. Beg, H.D. Politzer and P. Ramond, Phys. Rev. Lett. 43, (1979) 1701.

[15] M.A.B. Beg (private communication).

QUARKS IN LIGHT BARYONS*

Nathan Isgur

University of Toronto, Toronto, Ontario, Canada and

Gabriel Karl

University of Guelph, Guelph, Ontario, Canada
(presented by Gabriel Karl)

ABSTRACT: Recent work on quark models for baryons based on QCD is reviewed briefly.

I. INTRODUCTION, ASSUMPTIONS AND MODEL FOR BARYONS

It is now optimistically believed that all low energy particle physics phenomena, say below 5 GeV, are basically understood in the same rough sense in which Atomic Physics was understood in 1930. In the particular case of the hadron sector, Quantum Chromodynamics (QCD) is believed to be the basic theory. It is not certain as yet whether this belief is warranted, since exact results based rigorously on QCD are in short supply. In the absence of such exact results one has to make some guesses and hope that these guesses will be substantiated later from QCD. The work on baryons discussed here is mainly in this fashion of guess work mainly about the nature of interquark interactions.

---

*Work supported by N.S.E.R.C. Ottawa

There has been a great deal of recent work on hadrons primarily under the stimulation of the experimental discovery of J/ψ, ϒ and related particles. The first theoretical wave of attack was directed at explaining the charmonium and other heavies. The principles developed for charmonia have been applied also to lighter hadrons, mesons and baryons, even though the use of nonrelativistic models is much harder to justify in these cases. Therefore we are discussing "post-charmonium" baryon models.

Relative to the pioneering work of Greenberg, Dalitz and their coworkers on baryons, there are new physical assumptions which restrict the range of possible interactions between quarks. These assumptions refer to properties of the long range 'confinement' and short range quark quark potential. The long range interaction is assumed to be spin independent and more importantly flavor independent. This assumption is similar to the universality of the Coulomb interaction in positronium and muonium. In the hadronic sector, the more familiar case is the assumption of a universal potential in J/ψ (charmonium) and in ϒ (bottomonium), which is based in turn on the assumption that all quarks are color triplets. We similarly assume that a single (Lorentz scalar) potential function between quarks describes all baryons, irrespective of the number of strange quarks they contain. We take for this function a perturbed harmonic potential, with the perturbation treated in first order only. This choice is made for computational convenience.

The short range interaction is assumed appropriate to massless vector exchange, very similar, except for overall normalization to single photon exchange. The spin dependent part of this interaction, the so called 'hyperfine interaction' between quarks, first advocated by de Rujula et al. plays an important role also in light hadrons. The choice of (Lorentz) scalar confining potential has consequences also for the spin dependent interactions -- it leads to a large cancellation of spin-orbit forces -- as emphasized especially by Schnitzer.

The Hamiltonian for baryons reviewed here is

$$H = \sum_{i=1}^{3} m_i + \sum \frac{\vec{P}_i^2}{2m_i} + \sum_{i<j} V_{ij}^{conf}(r_{ij}) + \sum_{i<j} V_{hyp}^{ij} ,$$

$$V_{ij}^{conf}(r_{ij}) = \frac{1}{2} k\, r_{ij}^2 + \bar{V}(r_{ij}) ,$$

$$V_{hyp}^{ij} = \frac{2\alpha_s}{m_i m_j} \left[\frac{8\pi}{3} \vec{S}_i \cdot \vec{S}_j\, \delta^3(\vec{r}_{ij}) + \frac{1}{r_{ij}^3} (3\, \vec{S}_i \cdot \hat{r}_{ij} \vec{S}_j \cdot \hat{r}_{ij} - \vec{S}_i \cdot \vec{S}_j) \right] .$$

The idea of universality, or flavor independence consists in the confinement potential V_{ij}^{conf} being the same for any choice of quark masses.

II. DISCUSSION OF SOME SPECTROSCOPIC RESULTS

Given the choice of Hamiltonian it is a straightforward matter to find its eigenstates and corresponding eigenvalues in the various sectors - nonstrange, strangeness minus one, minus two etc. These results have been discussed in the literature and are too lengthy to describe in great detail here. Instead we only emphasize some general features of these results. These are:

i) Mode splitting in excited hyperons. This phenomenon corresponds to a shift in frequency dependent on the type of mode: if the strange quark is excited in a mode the frequency is lower than in a similar mode with the strange quark at rest. The simplest example is the splitting between $\Lambda 5/2^-$ and $\Sigma 5/2^-$, which arises as a consequence of this splitting. In the $\Lambda 5/2^-$ it is the nonstrange quarks which are excited, while in $\Sigma 5/2^-$ it is the strange quark which is excited, thus accounting for most of the splitting between these two states. This phenomena is somewhat similar to isotope shifts in molecules which arise because the forces between nuclei are the same in two different isotopes and only the masses change between them.

The selection of modes which depend on which quark is moving, strange or nonstrange is a particularly simple way of SU_3 breaking, which is strongly supported by the data. In the case of the P-wave 70 plet the basis chosen is of the type $|1\rangle \pm |8\rangle$ or $|10\rangle \pm |8\rangle$; in other words the axes are at 45^0 relative to the SU_3 axes. To take one simple example, the well known analysis of Hey, Litchfield and Cashmore, gave as the empirical composition of the $\Lambda 1/2^-(1405)$:

$$\text{(A)} \qquad |\Lambda(1405)\rangle = .85|^2 1\rangle + .46|^2 8\rangle + .25|^4 8\rangle \quad .$$

If we reexpress this state in terms of the rotated basis

$$|^2\rho\rangle = \frac{1}{\sqrt{2}}\left(|^2 1\rangle - |^2 8\rangle\right) \quad ,$$

$$|^2\lambda\rangle = \frac{1}{\sqrt{2}}\left(|^2 1\rangle + |^2 8\rangle\right) \quad ,$$

$$|^4\rho\rangle = |^4 8\rangle \quad ,$$

we find

$$\text{(B)} \qquad |\Lambda(1405)\rangle = .93|^2\lambda\rangle + .27|^2\rho\rangle + .25|^4\rho\rangle \quad ,$$

which shows that this state is almost pure $^2\lambda$ as one might also expect from its low mass. The comparison of (A) with (B) shows how much more convenient it is to use the ρ, λ basis to express experimental findings when compared to the SU_3 basis (A) which is far from the physical states. Similar observations can be made in the case of the positive parity excited baryons as well as the other strangeness sector $S = -2$.

ii) Spin dependent doublet mixings. A glance at the state (A) (or its other expression (B)) shows that there is a relatively small but nonzero mixing between spin doublet and spin quartet states. This is also the case with states in the nonstrange sector where we do not have to worry about mode splitting due to a strange quark. It is relatively easy to show that the hyperfine interaction accounts quantitatively for the observed doublet-quartet mixing in excited baryons. This mixing accounts for the decays of P-wave baryons which constituted a problem for quark models until the quark hyperfine interactions were taken into account.

III. SU_6 BREAKING

Other consequences of quark hyperfine interactions are related to the breaking of SU_6. This involves e.g. mixing into the ground state baryons (which have a symmetric orbital wavefunction) of excited baryons with a 'mixed' orbital wavefunction. The consequences are quite remarkable.

For example, there is a theorem of Fishbane et al. which says that the neutron charge radius must vanish as long as the neutron is a 56-plet. Hyperfine interactions, as shown by Ellis et al. and by Isgur, give rise to a nonzero charge radius, which has the right sign and size if the hyperfine interactions account for Δ-N splitting. This is easy to see intuitively - the Δ is heavier than the nucleon - therefore the hyperfine interactions must be repulsive between quarks with parallel spin. But this leads to a net repulsion between the two down quarks in the neutron (which have parallel spins) and to a negative charge radius - as observed experimentally.

Other consequences of mixing a 70-plet into the ground state baryons are the violation of the Moorhouse selection rule which forbids the photo decay of ^4P excited protons into the proton. With a 70 plet component in the proton these decays are allowed and the amplitude and helicity structure are in reasonable agreement with experiment. There is a similar selection rule by Faiman and Plane involving the $\bar{K}N$ decay of ^4P strangeness minus one hyperons whose observed violation is in reasonable agreement with the computation based on 70 plet mixing in the nucleon as argued by Koniuk et al.

Finally one can actually compute the decay amplitudes of a large set of excited nucleons, say the n=1 and n=2 configurations. This involves very laborious computations, which were recently completed by Isgur and Koniuk. A very noteworthy feature of these computations is that many states do decouple from the elastic channel as a result of the mixings due to hyperfine interactions. These states do account for so called 'missing' states in the quark model - which were not found in partial wave analyses of elastic scattering. Thus hyperfine interactions help tremendously in the comparison between quark model computations and experimental data.

REFERENCES

For a more complete set of references see:

Isgur, Nathan, lectures at the XVIth International School of Subnuclear Physics, Erice, Italy (1978) to appear.

Karl, Gabriel, Proceedings of the XIXth International Conference on High Energy Physics Tokyo (1978), edited by S. Homma, M. Kawaguchi and H. Miyazawa (Phys. Soc. of Japan, Tokyo 1979) p. 135.

Hey, A.J.G., Lecture at the European Physical Society Meeting in Geneva 1979 (Southampton preprint).

<u>More recent work</u>:

SU_6 Breaking: N. Isgur et al. Phys. Rev. Letters <u>41</u>, 1269 (1978).

Baryon Decays: R. Koniuk and N. Isgur, Toronto preprints (1979).

Ground State Baryons: N. Isgur and G. Karl, Phys. Rev. <u>D20</u>, 1191 (1979).

Charmed Baryons: L.A. Copley et al. Phys. Rev. <u>D19</u>, 768 (1979).

COLOR VAN DER WAALS FORCES?

O.W. Greenberg*

University of Maryland

College Park, Maryland 20742

It is very good to follow Gabriel Karl's excellent talk on new developments in the quark model of hadron spectroscopy, because a good deal of what I say is based upon the motivation of using the potential model that is so successful in baryon spectroscopy as well as in charmonium, upsilonium, and even light meson spectroscopy. On the other hand, in a way it is difficult to follow Gabriel Karl, because it is hard to match the charm and wit with which he delivered his talk.

I consider the question, "If quarks are permanently confined, how can hadrons separate without anomalous long range interactions which we would call color Van der Waals forces?"[1] By Van der Waals forces I mean inverse power forces analogous to the Van der Waals forces which occur, due to the electromagnetism, between atoms. I specifically do not use the phrase Van der Waals forces to refer to the exponentially decreasing meson-exchange mediated Yukawa interactions, even though these are also a residual effect of the colored gluon mediated interaction between quarks and antiquarks. Much of

*Supported in part by the National Science Foundation.

what I say is based on joint work with Manoj Banerjee and with Jarmo Hietarinta.

My basic starting assumptions are that permanent quark (and color) confinement is valid in hadronic physics; that it is worthwhile to see how far one can go assuming that only the quark degrees of freedom suffice for the description of hadrons at low energy; and that the two-body potential, which as I just remarked works well in baryon and meson spectroscopy, can be used also to describe systems of two or more hadrons. Later we will find that at least one of these assumptions must be modified.

First I will give the results of the two-body potential model applied to systems of two or more hadrons. The discussion can be divided into the abelian (or U(1)) and non-abelian (or $SU(n)_{color}$) cases, where for the non-abelian case, the real world corresponds to n=3. For the abelian case I will discuss only the formation of mesons. For the non-abelian case, both mesons and baryons can be formed; in this latter case I will usually discuss systems of two or more baryons where the constraints of the generalized Pauli principle for identical quarks are most restrictive. I will suppress the spin and flavor degrees of freedom of quarks. (As a parenthetical remark I point out that color-singlet exotic states occur in the models that I will discuss just as they occur in the usual description of hadrons. I will not discuss color-singlet exotic states further.)

I will assume that the confining potential grows as r^{α}, $r \to \infty$. For such growing potentials, I will ask several questions:

(1) Is the Hamiltonian H bounded from below?

If (1) is answered by yes, then

(2) Do separated hadron states exist?

If (1) and (2) are answered by yes, then

(3) What is the residual potential between neutral (for the abelian case) or color-singlet (for the non-abelian case) hadrons?

(4) Is it ever

$$V \sim r^{-1}\exp(-mr)? \quad ,$$

as we expect it should be for the lightest-mass allowed t-channel hadron exchange which gives the longest range Yukawa potential.

I will also make some remarks about what perturbative quantum chromodynamics (QCD) says for our problem. Of course I would prefer to give the results of nonperturbative QCD; however, such results are, unfortunately, not presently available.

Now I proceed to the detailed discussion of abelian models for mesons. Let the quark coordinates be x_i and the antiquark coordinates be y_i, where in general I will suppress the vector symbol over the x_i's and y_i's. For permanent confinement of quarks and antiquarks in mesons, I assume that the potential has the form

$$V_{q\bar{q}}(r) \to kr^{\alpha}, \quad r = |x-y| \to \infty \ . \tag{1}$$

In the later discussion, I will assume the potential actually has exactly its asymptotic form for all distances r. In order to cancel the most rapidly growing term, r^{α}, between the neutral clusters, I require

$$V_{qq} = V_{\bar{q}\bar{q}} = -kr^{\alpha}, \ r = |x_i - x_j| \text{ or } r = |y_i - y_j| \ . \tag{2}$$

For $\alpha > 2$, $V \to -\infty$, and H is unbounded below, even if $N_q = N_{\bar{q}}$. For example, consider the configuration shown in Fig. 1 where quarks are across one diagonal of a square of side a and antiquarks are across the other diagonal. In this case, the sum of all the

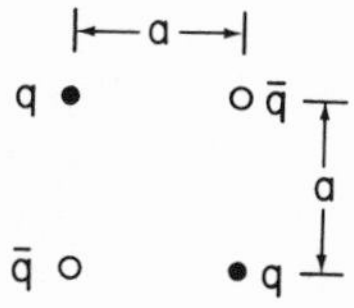

Fig. 1: Configuration of two quarks and two antiquarks for which the total potential $V \to -\infty$ as the scale $a \to \infty$, when the potentials $V_{q\bar{q}}$, V_{qq}, and $V_{\bar{q}\bar{q}}$ go as r^{α} with $\alpha > 2$.

two-body potential terms is

$$V = K[4a^{\alpha}-2(\sqrt{2}\ a)^{\alpha}] = (2^2-2^{\frac{\alpha+2}{2}})\ ka^{\alpha} \rightarrow -\infty,\ a \rightarrow \infty,\ \text{for } \alpha > 2. \tag{3}$$

For $\alpha=2$, $V \rightarrow -\infty$, and H is unbounded below, if $N_q \neq N_{\bar{q}}$. For example, for the case of two quarks and one antiquark on a line, with the antiquark halfway between the quarks as shown in Fig. 2, the total potential energy is

$$V = k[2a^2 - (2a)^2] = -2ka^2 \rightarrow -\infty,\ a \rightarrow \infty\ . \tag{4}$$

For the harmonic case, $\alpha=2$, with the number of quarks and antiquarks equal, $N_q = N_{\bar{q}} = N$, the sum of all the two-body potential terms is

$$V = k[\sum_{i,j} |x_i-y_j|^2 - \sum_{i<j} (|x_i-x_j|^2 + |y_i-y_j|^2)] = k[\sum_j (x_i-y_i)]^2\ , \tag{5}$$

where the second equality follows as an algebraic identity. For this case, the Schrödinger equation (assuming equal mass quarks and antiquarks) is

$$[-(2m)^{-1} \sum_1^N (\nabla^2_{x_i} + \nabla^2_{y_i}) + k\rho^2]\psi = E\psi\ , \tag{6}$$

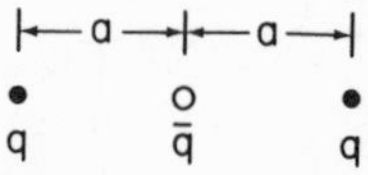

Fig. 2: Configuration of two quarks and an antiquark for which the total potential $V \rightarrow -\infty$ as the scale $a \rightarrow \infty$, when the potentials are harmonic ($\alpha=2$).

where

$$\rho = \sum_{1}^{N} (x_i - y_i) \quad . \tag{7}$$

In addition to the coordinate ρ, we also introduce the center-of-mass coordinate, X, and 3N-2 other traceless coordinates orthogonal to both the center-of-mass coordinate and to ρ, which we call r_λ. The general eigenstate of this Schrödinger equation has the form

$$\psi = H_{n\ell m}(\rho) \exp\left[-c\rho^2 + iK\cdot X + i\Sigma\, k_\lambda\cdot r_\lambda\right] \quad , \tag{8}$$

where H is a three-dimensional Hermite polyomial, K is the total momentum of the system, and the k_λ are the momenta conjugate to the coordinates r_λ. Here the constant c and the ground state energy E_o have the values

$$c = \left[(mk)/(4N)\right]^{1/2} \text{ and } E_o = 3\left[NK/m\right]^{1/2} \quad . \tag{9}$$

The generic eigenstate for an N-meson system would have the form

$$\psi_{N\text{-mesons}} = \left\{ \prod_{1}^{N} H_{n_i \ell_i m_i}(x_i - y_i)\exp\left[-c'(x_i - y_i)^2\right]\right\} \exp\, iK\cdot X \; , \tag{10}$$

with

$$c' = (mk)^{1/2}/2, \text{ and } E_o' = 3N\left[k/m\right]^{1/2} \quad . \tag{11}$$

Comparing the exact solution with the solution for N separated mesons, we see that mesons are not formed in the many-particle system with harmonic interactions. To put it another way, the residual interactions between separated mesons dissolve the mesons. Note that the energy of what should be an N meson system goes as $N^{1/2}$ rather than as N.

It is worthwhile to look at this problem in another way which does not require the exact solution of the Schrödinger equation. The potential for two separated neutral clusters for the harmonic case has the form

$$V = V_I + V_{II} + 2k\, \rho_I \cdot \rho_{II}, \text{ independent of the separation } a\,. \tag{12}$$

Thus a constant residual potential between separated neutral clusters prevents meson formation. This example will be useful for us in the non-abelian case when in general it is not possible to solve the Schrödinger equation exactly.

For $\alpha < 2$, it seems likely that one can use perturbation theory to calculate the potential between separated neutral clusters. The residual potential has the form

$$V - V_I - V_{II} = k\alpha\, |a|^{\alpha-2} \sum_i \sum_j \Big[(x_i - y_i)\cdot(x_j - y_j) + \frac{1}{2}\left(\frac{\alpha}{2} - 1\right)(\hat{a}\cdot(x_i - y_i))(\hat{a}\cdot(x_j - y_j))\Big]\,. \tag{13}$$

Here we have assumed that the number of quarks equals the number of antiquarks in the total system as well as in each of the neutral clusters. What is important about this equation is that

$$V - V_I - V_{II} \sim |a|^{\alpha-2}, \text{ and odd in } x_i - y_i \text{ and } x_j - y_j\,. \tag{14}$$

Note that in addition to the power behavior of the residual potential terms, there is a quantity linear in the relative coordinates inside each of the clusters. Therefore, by parity conservation, the first order perturbation vanishes. The second order perturbation goes as

$$|a|^{2\alpha-4}\,. \tag{15}$$

If we guess, by analogy with the Van der Waals potential for the Coulomb interaction, that retardation will produce an additional 1/a factor, then we expect

$$|a|^{2\alpha-5} \quad . \tag{16}$$

Thus for $\alpha=1$ we find a^{-1}; and for $\alpha=-1$ we find a^{-7}. Of course for no case with a growing potential do we get an exponential decrease. The moral of our discussion so far is that confining two-body potentials inside hadrons will produce Van der Waals analog interactions between separated hadrons which decrease as an inverse power of the separation between the neutral clusters. We never get an exponentially decreasing residual interaction.

For non-abelian cases, it is difficult to give as detailed a discussion as we have just done, because in general we get matrix systems of Schrödinger equations which are not exactly solvable even for the harmonic case. Nonetheless it is easy to see that pathologies still occur for the harmonic case $\alpha=2$. For example, for three quarks in an $SU(2)_{\text{color}}$ model in which a two-quark system in an isocolor singlet is a baryon, we find that for total color 3/2 states, the potential $\rightarrow -\infty$, and $H \rightarrow -\infty$. For the isocolor 1/2 case, which corresponds to a "baryon" together with an additional quark, we find that the potential is positive definite and thus also $H \geq 0$. In this case we can reduce the system of Schrödinger equations to a 2 x 2 matrix system. For Fermi quarks, there is an exact solution in which all three quarks are free, i.e. the potential term has entirely disappeared. For this solution the third quark has dissolved the "baryon." For Bose quarks, there is an exact solution in which all three quarks are bound by interquarks, there is an exact solution in which all three quarks are bound by interquark harmonic potentials having the same strength as the original two-body potential, and thus there is an "exotic baryon" which has the color quantum numbers of the single quark.

Although we cannot solve the Schrödinger equations exactly in the harmonic case, we can discuss the residual potential when a color-singlet system of quarks is separated into two color-singlet clusters. We assume a potential of the form

$$V = -k \sum_{i<j} \lambda_i^\alpha \lambda_j^\alpha (x_i - x_j)^2 \text{ (sum on } \alpha \text{ understood)}$$

$$= k\{[\sum_i \lambda_i^\alpha x_i]^2 - \frac{1}{2}[\sum_i \lambda_i^\alpha x_i^2, \sum_j \lambda_j^\alpha]_+\} \quad , \qquad (17)$$

where

$$\sum_j \lambda_j^\alpha = 2 I^\alpha \quad . \qquad (18)$$

For overall color singlet states, the matrix element of the potential assumes a simple form

$$\langle I_c^\alpha = 0|V|I_c^\alpha = 0\rangle = k\langle I_c^\alpha = 0|(\vec{\mathcal{D}}^\alpha)^2 \, I_c^\alpha = 0\rangle \quad , \qquad (19)$$

where

$$\vec{\mathcal{D}}^\alpha = \sum_i \lambda_i^\alpha \vec{x}_i \quad . \qquad (20)$$

If we separate the total color-singlet state into two color-singlet clusters with separation a, the residual potential is

$$V - V_I - V_{II} = 2k\vec{\mathcal{D}}_I^\alpha \cdot \vec{\mathcal{D}}_{II}^\alpha \text{ , independent of a.} \qquad (21)$$

By analogy with the abelian case, we expect that this separation-independent residual interaction will prevent formation of baryons. On the other hand, just as in the abelian case, we expect that for $\alpha < 2$ hadrons will form and the color Van der Waals potential will go as $a^{-2\alpha-4}$, or $a^{2\alpha-5}$ with retardation. Once again we find that there is no exponential decrease.

Perturbative QCD corresponds to the case $\alpha = -1$, which is the same power as the Coulomb potential in electrodynamics, and leads to a residual interaction between hadrons going as a^{-7}. Of course, if the nonperturbative solution of QCD has confinement then such long-ranged Van der Waals interactions would not occur. If confinement occurs, and if, in addition, the lowest-mass hadronic state has positive mass, then an exponential decrease would occur.

For local quantum field theories with a positive metric in Hilbert space and manifest Lorentz invariance, inverse-power potentials imply a mass spectrum going down to zero. Since there is no experimental evidence that the mass spectrum of hadrons reaches zero, the solution to hadronic physics cannot have inverse-power potentials. One must, however, inject a note of caution since this theorem has not been proved, to my knowledge, in a context relevant to gauge theory, in which one cannot simultaneously have a positive metric in Hilbert space and manifest Lorentz invariance.

Feinberg and Sucher[2] have recently done a comprehensive analysis of experimental upper limits on inverse power residual interactions between hadrons with the result that low inverse powers, such as one to three, are strongly ruled out by present data; however, the inverse power seven which would occur in perturbative QCD is not very constrained by present experimental data.

It is also possible that in fact the hadronic spectrum does reach zero even though no zero-mass particles actually exist. This possibility was discussed by Feinberg and Sucher and will be studied further by Hiller and Sucher.

If color Van der Waals potentials are indeed absent, then quark degrees of freedom alone do not suffice to describe systems of several hadrons. Bag models do avoid color van der Waals interactions between hadrons because in bag models the color gluons never leave a bag and therefore do no cross between the separated hadrons. Nonetheless, I do not consider bag models to be in the standard framework of quantum field theory, and therefore I think it worthwhile to find another way of avoiding color Van der Waals interactions.

Hietarinta and I studied the color Van der Waals problem and attempted to find a minimal modification of the potential model which would allow the elimination of long-ranged interactions between separated hadrons. We started from the point of view that some analog of the string/vortex gluon degree of freedom must be incorporated into the model in order to provide a way in which the quarks (and antiquarks) can know when they are in the same hadron, and to arrange that the confining potential acts only between the quarks in a given hadron. We found both a first-quantized scheme[3] and a second-quantized scheme[4] to accomplish this. The first-quantized scheme is published in Physics Letters, and I will not describe it further here except to say that several of the conclusions of the first-quantized scheme also hold for the second-quantized one:

(1) The results for a single hadron are the same as the potential model.

(2) There is a modification of the Pauli principle for quarks, although Fermi hadrons obey the usual Pauli principle and Bose hadrons obey the usual Bose statistics. The modification of the Pauli principle for quarks occurs because quarks are connected to other quarks or antiquarks by strings. When one interchanges a quark in one hadron with a quark or antiquark in a different hadron, the strings go with the quarks or antiquarks, and thus the state after exchange is distinguishable from the original state. Therefore the Pauli principle in its usual form does not hold for the exchange of quarks between different hadrons, although it does hold for permutations of quarks inside a given hadron or for permutations of identical baryons. In the latter case all the quarks in each baryon are exchanged together.

(3) There is also a new degeneracy of states, because states in which identical quarks have been exchanged, but in which dif-

ferent string configurations occur, are linearly independent. In this connection, it is worth pointing out the contrast between the string picture of quarks and the situation for covalent bonds in chemistry. The covalent bonds in chemistry occur because the energy is lower for electron pairs in the spin-zero state than for pairs in the spin-one state. However the interaction energy associated with a pair of electron spins in the spin-zero state spreads among other pairs of electrons, because the number of possible pairs (covalent bonds) is greater than the number of linearly independent states. This spreading of the interaction between electron pairs does not occur for the quark model with strings, which is thus a more saturated form of interaction than the covalent bonds in chemistry.

I now describe the second-quantized scheme to avoid color Van der Waals potentials between separated hadrons which Hietarinta and I found. I will carry out the discussion in the context of a nonrelativistic model in which the usual annihilation and creation operators for quarks and antiquarks,

$$q(x),\ q^{\dagger}(x),\ \bar{q}(y),\ \bar{q}^{\dagger}(y) \quad , \tag{22}$$

represent the degrees of freedom carried by quarks and antiquarks, and new operators which we call "link" operators

$$L(x,y)\ ,\ L^{\dagger}(x,y) \tag{23}$$

represent the string/vortex degree of freedom. I will carry out most of the discussion for the abelian case. The non-abelian case can be treated in an analogous manner. A single meson is created using these operators in a way similar to the usual construction for a single meson in the naive quark model, except that a link operator is introduced together with the quark and antiquark creation operators:

$$|M(p)\rangle = \int d^3x d^3y (\exp ip \cdot \frac{x+y}{2})\ \phi(x-y) q^\dagger(x) L^\dagger(x,y) \bar{q}^\dagger(y)|0\rangle \quad . \tag{24}$$

Here as usual, ϕ is the meson wave function, and p is the total meson momentum. We assume that the link operators have the commutator

$$[L(x,y),\ L^\dagger(x',y')] = \delta_K(x-x')\ \delta_K(y-y') \ , \tag{25}$$

where

$$\delta_K(x) = \begin{cases} 1, x=0 \\ 0, x\neq 0 \end{cases} \ , \tag{26}$$

and

$$[L,L] = 0,\ [L^\dagger,\ L^\dagger] = 0 \ . \tag{27}$$

There is a new and peculiar character to the link operator commutator: it involves the "Kronecker" δ function rather than the Dirac δ function. In an integration over the argument without an accompanying Dirac δ function, the Kronecker δ function would not contribute; however in our calculations the Kronecker δ function can contribute because it is accompanied by a Dirac δ function coming from contractions of the quark and antiquark operators. The basic equations are

$$\delta_K(x)\ \delta(x) = \delta(x) \tag{28}$$

and

$$\delta_K(x)\ \delta(x') = 0,\ x \neq x' \ . \tag{29}$$

In order to make a consistent scheme, we also need analogs of number operators for the link operators. As a motivation for the commutators of the number operators, we use the equations

$$N(x,x') \sim \int L^{\dagger}(x,y)\ L(x',y)d^3y \quad , \tag{30}$$

and

$$\bar{N}(y,y') \sim \int L^{\dagger}(x,y)\ L(x,y')d^3x\ ; \tag{31}$$

however we do not literally use these formulas. Our motivation leads us to a closed set of commutators among the operators N, $\bar{N}$, L and $L^{\dagger}$:

$$[N(x,x'),\ L^{\dagger}(x_1,y)] = \delta_K(x'-x_1)\ L^{\dagger}(x,y) \quad , \tag{32}$$

$$[N(x,x'),\ N(x_1,x_1')] = \delta_K(x'-x_1)N(x,x_1')-\delta_K(x-x_1')N(x_1,x') \tag{33}$$

$$[\bar{N}(y,y'),\ L^{\dagger}(x,y_1)] = \delta_K(y'-y_1)\ L^{\dagger}(x,y) \quad , \tag{34}$$

$$[\bar{N}(y,y'),\ \bar{N}(y_1,y_1')] = \delta_K(y'-y_1)\ \bar{N}(y,y_1')-\delta_K(y-y_1')\bar{N}(y_1,y') \quad , \tag{35}$$

$$[N(x,x'),\ \bar{N}(y,y')] = 0. \tag{36}$$

We have checked that all Jacobi identities are satisfied for these commutators, and thus these commutators are a consistent set for the operators we are using. In addition we assume that the quark operators commute with the link operators

$$[q^{\dagger}(x),\ L^{\dagger}(x',y)] = 0,\ \text{etc.}, \tag{37}$$

and we assume that the quark and antiquark annihilators annihilate the vacuum as does the link operator and also the number operators N and $\bar{N}$,

$$q(x)|0\rangle = 0,\ L(x,y)|0\rangle,\ N(x,x')|0\rangle = 0,\ \text{etc.} \tag{38}$$

With this set of commutators and vacuum state conditions, we can calculate all matrix elements of these operators. It is straightforward to see that the inner product among single-particle meson states gives the usual result where in particular we use the identity Eq. (28,29) above. For two-particle matrix elements, such as

$$\langle M_1'(p_1')\ M_2'(p_2')|M_1(p_1)M_2(p_2)\rangle \quad , \tag{39}$$

we find that two types of matrix elements involving the link annihilation and creation operators occur after the quark contradictions have been done: These are

$$\langle 0|L(x_1,y_1\ L(x_2,y_2)\ L^\dagger(x_1,y_1)\ L^\dagger(x_2,y_2)|0\rangle = 1, \tag{40}$$

and

$$\langle 0|L(x_1,y_1)\ L(x_2,y_2)\ L^\dagger(x_1,y_2)\ L^\dagger(x_2,y_1)|0\rangle = 0\ . \tag{41}$$

The link commutation relations make the first of these expressions unity, because both arguments of a link annihilator agree with both arguments of a link creator; however, the second of these vanishes, because this condition is not satisfied. The first matrix element corresponds to the case in which a quark-antiquark pair in a meson on one side of the inner product contracts with a quark-antiquark pair in a single meson on the other side of the inner product; while the second (vanishing) matrix element corresponds to the case where a quark-antiquark pair in a meson on one side of the inner product contracts with quarks and antiquarks in two different mesons on the other side of the inner product. The vanishing of this latter term corresponds to the elimination of the Van der Waals potentials.

Proper formulation of this model requires that the links and quarks remain together. To insure that this occurs it is necessary to include the number operators in the kinetic terms: for example, the momentum operators have the form

$$P^i = i\int d^3x \; ({}^1\nabla^i_x \; q^\dagger(x) \; N(x,x)) \; q(x) \; +$$

$$+ \; i\int d^3y \; ({}^1\nabla^i_y \; \bar{q}(y) \; N(y,y))\bar{q}(y) \quad , \tag{42}$$

where ${}^1\nabla$ acts on the first argument of N or $\overline{N}$ (in addition to acting on q or $\bar{q}$).

In this model, the states and operator densities are locally gauge invariant and integrated operators are gauge invariant under the set of local gauge transformations:

$$q^\dagger(x) \rightarrow q^\dagger(x) \; U^\dagger(x) \quad , \tag{43}$$

$$\bar{q}^\dagger(x) \rightarrow \bar{q}^\dagger(y) \; U^T(y) \quad , \tag{44}$$

$$L^\dagger(x,y) \rightarrow U(x) \; L^\dagger(x,y) \; U^\dagger(y) \quad , \tag{45}$$

$$N(x,x') \rightarrow U(x) \; N(x,x') \; U^\dagger(x') \quad , \tag{46}$$

and

$$\overline{N}(y,y') \rightarrow U^{\dagger T}(y) \; \overline{N}(y,y') \; U^T(y') \; , \tag{47}$$

where T stands for transpose and is relevant for non-abelian models. The Hamiltonian for the simplest model which we considered in which only the confining potential acts and there is no creation or annihilation of quarks or links has the form

$$H = -(2m)^{-1} \int d^3x ({}^1\nabla^2_x q^\dagger(x) \; N(x,x)) \; q(x)$$

$$-(2m)^{-1} \int d^3y \; ({}^1\nabla^2_y \; \bar{q}^\dagger(y) \; \overline{N}(y,y)) \; \bar{q}(y)$$

$$+ \int d^3x \; d^3y \; q^\dagger(x) \; L^\dagger(x,y) \; \bar{q}^\dagger(y) \; N(x-y) \; \bar{q}(y) \; L(x,y) \; q(x) \; . \tag{48}$$

Note that the expression $q^\dagger L^\dagger \bar{q}^\dagger$ acting on the vaccum creates a superposition of single meson states:

$$q^\dagger(x)\ L^\dagger(x,y)\ \bar{q}^\dagger(y)|0\rangle = \sum_j \int d^3w\ \phi_j(\tfrac{1}{2}(x-y)-w,x-y)\ M_j^\dagger(w)|0\rangle \ . \tag{49}$$

We have now succeeded in eliminating Van der Waals forces and have a model in which quarks and antiquarks are permanently confined in mesons and the mesons separate freely, indeed, have no interactions between them.

The non-abelian case is similar to the abelian case just discussed except that the color degree of freedom must be taken into account. We do that by assigning a single $SU(3)_{color}$ index to the quark and antiquark operators and a pair of color indices to the link operators; for example, we replace $q^\dagger L^\dagger \bar{q}^\dagger$ by

$$q^{\dagger\alpha}(x) L^{\dagger}{}_{\alpha}{}^{\beta}(x,y)\ \bar{q}^{\dagger}{}_{\beta}(y) \quad . \tag{50}$$

It is easier to describe our calculations in diagrams than in formulas:

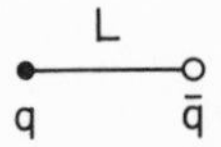

Fig. 3: Meson as quark-link-antiquark.

Figure 3 illustrates the form of a meson and Fig. 4 the form of a baryon.

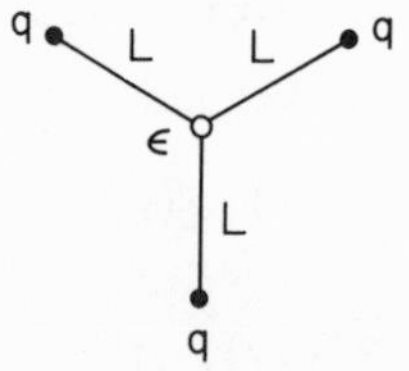

Fig. 4: Y configuration for a baryon.

For the baryon, following the work of people in the dual models, we have introduced the operator ε as a junction operator. The Y configuration which we introduce requires an extra degree of freedom, namely the degree of freedom at the junction; however, one can arrange to make excitation of the junction degree of freedom arbitrarily high in energy. Physically the junction degree of freedom may represent excitation of the gluon degree of freedom. The V configuration, Figure 5, like the Y configuration, is locally gauge invariant; however the V configuration has an extra degeneracy since one of the quarks, namely the quark at the junction of the V, is treated differently than the other two.

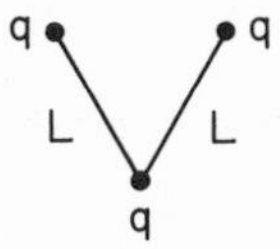

Fig. 5: V configuration for a baryon

The triangle configuration, Figure 6, can be made globally gauge invariant but cannot be made locally gauge invariant.

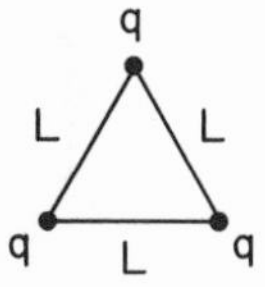

Fig. 6: Triangle configuration for a baryon.

Just as in the abelian case, color singlet exotics can occur.

Both the second-quantized scheme which we have just sketched, and the first-quantized scheme, allow the confining potential to have any growth; the restrictions concerning the rate of power growth which I discussed for the two-body potential model are no longer necessary. In addition, in both schemes the confining potential acts only in attractive channels; there is no need to have canceling repulsive potentials in other channels. In particular, for the relevant case of $SU(3)_{color}$ the confining interaction acts between two quarks in the 3* channel and a quark-antiquark pair in the singlet channel.

Interactions can be introduced via link breaking and link rearrangement: Again I indicate this via diagrams. Fig. 7 shows a link re-arrangement process corresponding to quasi-elastic meson scattering,

Fig. 7: Quasi-elastic meson-meson scattering via link re-arrangement.

and Fig. 8 shows a link-breaking process corresponding to meson-meson to three meson production.

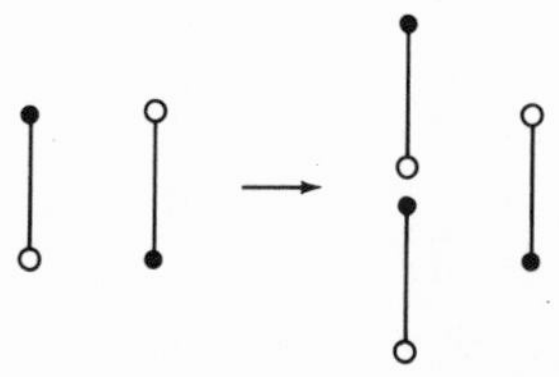

Fig. 8: Meson production via link breaking.

One can also introduce vacuum to $qL\bar{q}$ virtual processas as shown in Fig. 9.

Fig. 9: Vacuum to $qL\bar{q}$ virtual process.

Having accomplished our goal of eliminating color Van der Waals interactions between hadrons, it is important to find independent experimental predictions of these ideas. One possibility is to study the Iizuka-Okubo-Zweig rule; work in this direction is now being carried out by Achim Weidemann and myself. The modification of the Pauli principle and the counting of states should, in principle, also lead to experimental tests of these ideas; however at present it is not clear how to give sharp experimental tests of these modifications.

I want to conclude this talk by pointing out that something very similar to the link operators occurs in the large-N limit of SU(N) theories.[5] Consider relativistic SU(N) fields, and, in particular, consider a matrix element

$$\langle 0|\psi_{\alpha_1}^{(+)}\bar{\psi}^{\alpha_1(+)}\psi_{\alpha_2}^{(+)}\bar{\psi}^{\alpha_2(+)}\psi_{\beta_1}^{(-)}\bar{\psi}^{\beta_1(-)}\psi_{\beta_2}^{(-)}\bar{\psi}^{\beta_2(-)}|0\rangle, \tag{51}$$

which corresponds to the inner product between two-meson states. The evaluation of this matrix element gives two kinds of terms: one of which has the form

$$\delta^{\alpha_1}{}_{\beta_1}\,\delta_{\alpha_1}{}^{\beta_1}\,\delta^{\alpha_2}{}_{\beta_2}\,\delta_{\alpha_2}{}^{\beta_2} = N^2 \quad . \tag{52}$$

Terms of this form occur when the creation parts of ψ and $\bar{\psi}$ fields which are in an SU(N) singlet contract with annihilation parts which are also in an SU(N) singlet, and correspond to matrix elements in which the quark and antiquark in one meson contract with a quark and antiquark in another meson on the other side of the inner product. The terms of the other form,

$$\delta^{\alpha_1}{}_{\beta_2}\,\delta_{\alpha_2}{}^{\beta_2}\,\delta^{\alpha_2}{}_{\beta_1}\,\delta_{\alpha_1}{}^{\beta_1} = N \quad , \tag{53}$$

come from terms in which the creation parts which are an SU(N) singlet contract with ψ and $\bar{\psi}$ annihilation parts which are in two different SU(N) singlets on the other side of the matrix element, and correspond to matrix elements in which the quark and antiquark in a single meson on one side of the inner product contract with a quark and antiquark in two different mesons on the other side. The analogy with the situation for link operators is clear. In the SU(N) limit, for $N \to \infty$, the terms of order N are eliminated, and thus one recovers the link operators recipe. It is worth remarking, that if one defines new fields by the limiting matrix elements that only the products

$$N^{-1/2}\ \psi_\alpha \bar{\psi}^\alpha \to [\chi\bar{\chi}] \tag{54}$$

are defined in the large N limit; in particular, the field χ itself is not defined. Further, the product $[\chi\bar{\chi}]$ is non-associative.

Finally, in addition to the issue that we must find independent tests of the ideas which I mentioned above, there is another important open question concerning these ideas: is this scheme merely phenomenological, or can it be made into a relativistic field theory?

FOOTNOTES AND REFERENCES

1. More detailed discussion of this question together withcitations to the literature will appear in the proceedings of the XVIIth Winter School of Theoretical Physics at Karpacz.
2. G. Feinberg and J. Sucher, Phys. Rev. D20, 1717 (1979). I rely on this reference for citations to other relevant literature.
3. O.W. Greenberg and J. Hietarinta, Phys. Lett. 86B, 309 (1979).
4. O.W. Greenberg and J. Hietarinta, Brookhaven National Laboratory Report BNL-26463 (1979), being revised.
5. O.W. Greenberg, University of Maryland Technical Report, No. 80-027 (1979), being revised.

THE QUESTION OF PROTON STABILITY

SUMMARY

Maurice Goldhaber

Brookhaven National Laboratory

Upton, New York 11973

The talk dealt with some of the historical, experimental and theoretical facts concerning the question of proton stability. These have recently been reviewed[1] and will not be repeated in this summary. Earlier short reviews might also be consulted.[2,3] The present status of the theoretical predictions for the proton lifetime based on the SU(5) model is given in the report at this conference by R. Marciano.[4] This theory gives an upper limit for the proton lifetime: $\tau_p < 5 \times 10^{32}$ years (see also Ref. 5). Similar predictions follow from some theories based on larger groups which also contain $SU(3)^c \times SU(2) \times U(1)$.

It embarrasses me to contradict the sentiment expressed in the 1963 official BNL Christmas card:

SPEAK OF EXPERIMENTS IN THE PAST TENSE,
OF THEORIES IN THE PRESENT TENSE,
OF MACHINES IN THE FUTURE TENSE.

However, it is now common to talk of experiments in the future tense. Some experiments with an expected sensitivity which may be sufficient to check the theoretical predictions are now planned or underway. The experiments known to me are given in Table I. They are all to be carried out at sufficient depth to absorb a good deal of the cosmic muon background and with detectors varying from ∿10,000 tons to a few hundred tons. Preliminary results may be a year away.

TABLE I

P LIFETIME SEARCHES

IRVINE - MICHIGAN - BROOKHAVEN (Salt mine)

HARVARD - PURDUE - WISCONSIN (Silver mine)

PENNSYLVANIA (Gold mine)

MINNESOTA (Iron mine)

CERN - FRASCATI - MILANO - TURINO (Mont Blanc Tunnel)

SACLAY - ORSAY (Frejus Tunnel)

REFERENCES

1. M. Goldhaber, P. Langacker and R. Slansky, Science to be published.
2. F. Reines in "Unification of Elementary Forces and Gauge Theories", D.B. Cline and F.E. Mills, Editors, Harwood Academic Publishers, London-Chur 1977, p. 103.
3. M. Goldhaber, ibid, p. 531.
4. R. Marciano, p. 121 of this Conference Report.
5. T.J. Goldman and D.A. Ross (Cal Tech Preprint, 1980) give an upper limit of 2×10^{32}y.

A REVIEW OF CHARMED PARTICLE PRODUCTION IN HADRONIC COLLISIONS*

Stephen L. Olsen

The University of Rochester

Rochester, New York 14627

ABSTRACT

A review of recent experimental results on the production of Charmed Particles in hadronic collisions is reported. Comparisons with theoretical predictions are made. Results from Neutrino beam dump experiments at the CERN SPS and direct muon production at Fermilab indicate that central production of Charmed mesons occurs with a total cross section of 10 to 20 microbarns. Charmed mesons and Baryons have also been observed at the CERN ISR with cross sections that appear to be considerably larger. Some recent experimental limits on the production of B particles are also reviewed.

I. INTRODUCTION

Since the initial observation of the J/ψ[1,2] in November 1974, considerable effort has been expended in searching for the production of charmed particles in hadronic collisions. Initially this was motivated by the desire to establish the existence of charmed particles. Subsequently, after the observation of charm in e^+e^-

---

*Work supported in part by the U.S. Department of Energy under Contract NO. EY-76-C-02-3065.

annihilations[3] and neutrino reactions[4,5], the emphasis on this effort has been to study the nature of charm production in hadronic processes. The understanding of these processes is important for a number of reasons, among them;

1) Checks of QCD

Because the charmed quark is expected to be heavy, $1/m_c \sim$ .1 fermi, perturbative QCD calculations are expected to be valid. This is one of the few hadronic processes which QCD should be able to compute reliably. Since similar calculations are used to estimate Z^o and $W^{\pm}$ production at the proposed high energy pp and $\bar{p}p$ colliders at CERN, Fermilab and Brookhaven, tests of these theories have considerable practical significance.

2) Sources for ν_e and ν_τ

The decay modes $D \to K(K^*)e\nu_e$ and $F \to \tau\nu_\tau$ are possible sources for beams of electron neutrinos and τ neutrinos respectively. The design of experiments using beams such as these requires reliable information on the production rates of these particles[6].

3) Study of charmed particle properties

Recent indications are the $\sigma(pp \to \text{charm}) \sim 20\mu b$. If so, hadronic collisions may prove to be a valuable tool for the detailed studies of the properties of charmed particles. For example, the cross section for $e^+e^- \to \psi(3770)$ (the "charm factory") is about 9 nb[7]. Thus, yields at e^+e^- storage rings are typically no more than 300 charmed meson events per day. Comparable rates in hadronic experiments could be on the order of 100 per machine spill or 10^6/day. Exploiting these high rates will require well designed triggers for which a good understanding of charm production properties is essential[8].

4) Backgrounds to New Particle Searches

Many proposed new particles such as B mesons, Higgs particles etc., are expected to be sources for directly produced leptons. Experimental searches for these new particles using direct leptons as a signature will have direct leptons from charm decay as a background. An understanding of the character of charm production is needed to evaluate this background.

II. POSSIBLE PRODUCTION MECHANISMS

Perturbative QCD estimates of charm production have recently been reviewed by Halzen[8]. They correspond to the two different types of diagrams shown in Fig. 1. Figure 1(a) shows some typical single gluon contributions. These are the same as the diagrams expected for electromagnetic production of muon (electron) pairs with gluons in place of virtual photons. The strength of this contribution for 400 GeV pp collisions can be estimated by multiplying the total dimuon cross section for $M_{\mu\mu} \gtrsim 3.5$ GeV ($\sim$0.1 nb)[10] by $(\alpha_s/\alpha)^2$ where α_s is the QCD quark gluon coupling constant ($\sim$.2). This gives an estimate for the total cross section for charmed production of $\sim$.1 μb. Halzen emphasizes the importance of two gluon contributions expected from the diagrams in Fig. 1(b). As the energy increases these processes are expected to dominate.

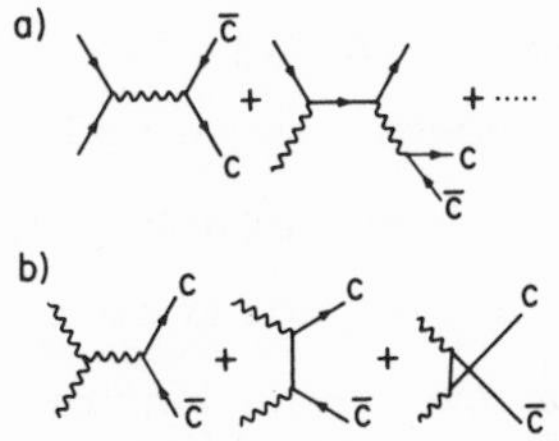

FIG. 1: (a) Single gluon contributions and (b) Two gluon contributions to heavy quark production.

Figure 2 shows the expected contribution to the charm production cross section from both types of processes. At 400 GeV, the two gluon processes dominate and a total cross section of $\sim$3 or 4 μb is expected for 400 GeV pp collisions. Included in Fig. 2 are similar estimates for B particle production. Here the one gluon diagrams still dominate at 400 GeV and a total $B\bar{B}$ production cross section of $\sim$10 nb is expected. Note that both the charm and B particle cross sections are expected to be about an order of magnitude higher at typical ISR energies (P_{lab} ∿ 2000 GeV).

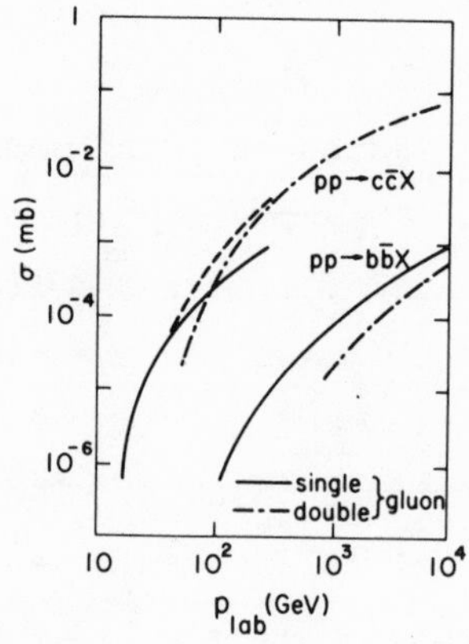

FIG. 2: Cross section estimates for c and b quark production (from ref. 9).

Another possible mechanism for charm production is the diffractive dissociation of one of the protons into a charmed baryon-anti charmed meson system. Quark line diagrams for possible diffractive processes are shown in Fig. 3. A rough estimate for the cross section for this process can be made by multiplying the cross section for Λ production by the ratio of the charmed component of the quark antiquark sea in the proton to the strange component. Results on same sign dimuon events in neutrino interactions indicate that

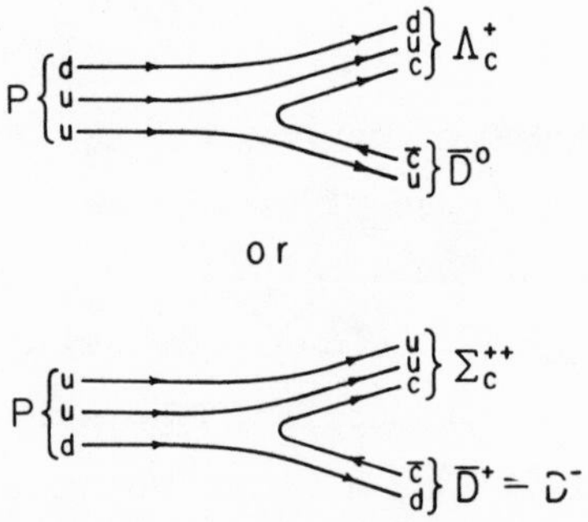

FIG. 3: Quark line diagrams for proton diffraction dissociation into charmed particle final states.

this ratio may be as high as 0.1. Since at the ISR, $\sigma_\Lambda \sim 3$ to 4 mb,[11] we expect a cross section for these processes which would be of the order of hundreds of microbarns.

A third mechanism for charm production has been suggested by Fritzsch and Streng[12]. They generalize the notion of photon vector dominance to gluon vector dominance. The basic diagram is indicated in Fig. 4. This model predicts σ charm $\sim 100\mu$ b at Fermilab/SPS energies and 500 μb at ISR energies. Furthermore, the model predicts substantial cross sections for B meson production, 5 μb at Fermilab/SPS and 20 μb at ISR.

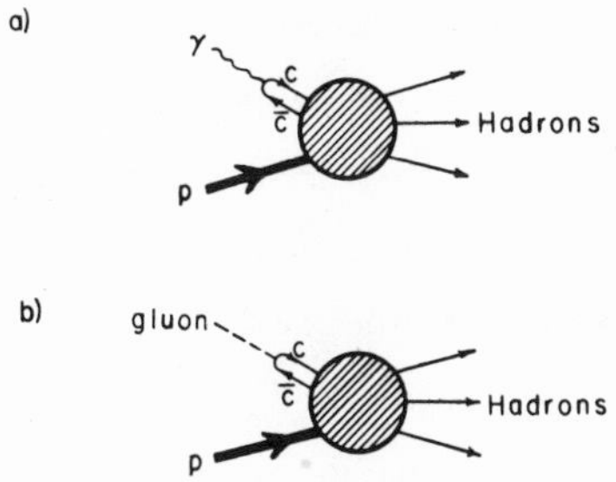

FIG. 4: (a) Photon Vector Dominance diagram and
(b) Gluon Vector Dominance diagram for charmed quark production.

III. TYPES OF EXPERIMENTS

A. ISR Experiments - Exclusive Channels[13]

Recent results from the ISR measured charm production detecting exclusive multihadron final states. In these measurements some particle identification is provided by threshold Cerenkov counters. In order to reduce combinatorial backgrounds final states with identified K^- or Λ^o's are chosen. The invariant masses of the possible track combinations are computed with the charmed particles showing up as peaks at the appropriate masses. Since, at the ISR, the detector is approximately in the CM system and these detectors do not have 4π acceptance, the parent charmed particles, if they are to be detected, must have considerable CM momentum if all the decay products are to be swept into the acceptance region. Thus these experiments are sensitive to production at non zero Feynman x (typically x $\sim$.5) and are thus mostly sensitive to diffractive production. These experiments identify precisely the charmed particle being produced. On the other hand the determination of a cross section is sensitive to a knowledge of decay branching ratios which are usually poorly determined.

B. SPS and Fermilab Inclusive Lepton Measurements

These experiments rely on the fact that charmed particles, because of their semileptonic decay modes and short lifetime, are a source of apparent directly produced leptons, either electrons and muons or neutrinos. The semileptonic branching ratio of the D mesons has been determined with some precision to be about 8% from storage ring measurements.[14] Recent measurements at the CERN SPS have inferred charmed meson production cross sections by measuring the rate of prompt neutrino production in beam dump experiments. Charmed particle production cross section measurements at Fermilab have been inferred from direct muon production experiments together with measurements of "missing" hadronic energy, indicating the presence of final state neutrinos which carry off a significant amount of energy.

i) Neutrino beam dump experiment[15]

The basic technique of these experiments is indicated in Fig. 5(a,b). While in a normal neutrino experiment (Fig. 5a) considerable space is left downstream of the production target to allow the produced π and K mesons to decay, in a beam dump experiment the hadron absorber is placed as close to the target as possible to prevent these same decays (Fig. 5b). Even then, because of the finite interaction length in these absorbers ($\sim$10 cm or so) some neutrinos from π and K decay are produced. In the most recent experiment a target of variable density was used which changed the effective length for π and K decays from 30 cm (expanded mode) to 10 cm (compressed mode). This enabled the extrapolation to zero decay length and a corresponding "prompt" neutrino rate. Also, since particles produced by the beam scraping on magnets etc., upstream of the target could have a considerable decay distance, great care was taken to reduce and monitor this scraping.

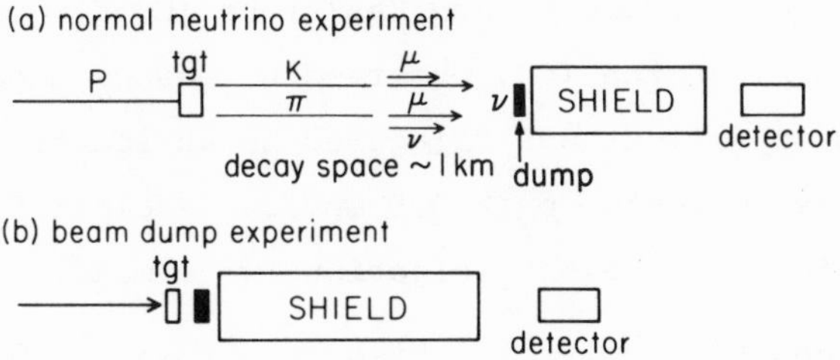

FIG. 5: (a) Schematic view of the normal neutrino experiment with a long drift space to allow π and K meson decays.
(b) A schematic view of a beam dump experiment where the drift space is eliminated.

ii) Direct muon measurements with missing energy[16]

In these experiments the proton beam is made to interact in a high precision hadron calorimeter which preceeds a large acceptance muon detector. Single muon production and double muon production with missing hadronic energy is measured and used to infer charm production cross sections. In these experiments π and K decay are also monitored by varying the target density and extrapolating to zero decay lengths. Another important background is electromagnetic dimuon production where one of the muons is not observed. For this reason a muon detector with large acceptance is used and the number of missed second muons determined by extrapolating the measured dimuon signal and by Monte Carlo calculations.

The inclusive lepton experiments do not measure the particular parent particle but since they can cover a large kinematic region and since the leptonic branching ratio for D mesons is reasonably well known (at least the average D^+D^o BR's)[14] their determination of the total charmed particle production cross section is more precise than the corresponding measurements of exclusive charmed particle production at the ISR.

IV. RECENT ISR RESULTS

A. CCHK Collaboration $D^+ \to K^*(890)\ \pi^+$ (ref. 12)

An experiment performed by a collaboration of CERN, College de France, Heidelberg and Karlsruhe (CCHK), using the Split Field Magnet (SFM) at the ISR, observed D^+ mesons decaying as $D^+ \to K^-\pi^+\pi^+$. This experiment was triggered by an identified K^- meson with a transverse momentum $p_T > 0.5$ GeV/c. A clear 5σ signal was seen when the following event selections were made:

1) The observed multiplicities (n_{obs}) was < 10.
2) A leading system was required in the opposite hemisphere.
3) The Feynman x of the K^- meson was > 0.3.
4) The p_T of the trigger K^- was compensated (to 0.2 GeV/c) in the same hemisphere.
5) The K^- and one π^+ had an effective mass of 890 ± 40 MeV, i.e., they form a K*(890).

To extract a cross section the signal had to be corrected for the branching ratio for $D^+ \to K^*\pi^+$, $K^* \to K^-\pi^+$, which is 2.6 ± 1% as measured in annihilation experiments[18]. This results in a differential cross-section $\frac{d\sigma}{dx}$ (x > 0.3) ≃ 55 μb with about 50% errors. If the production of D mesons is assumed to be uniform in x this gives a total cross section for D production of $\sigma_{D\bar{D}} \sim 110$ μb. If, on the other had, dσ/dx is assumed to go as $(1-x)^3$, then $\sigma_{D\bar{D}} \sim 1$ mb.

B. ACHMNR Collaboration, $\Lambda_c^+ \to K^- p\pi^+$ and $\Lambda^0\pi^+\pi^+\pi^-$ (ref. 19)

The collaboration of Aachen, CERN, Munich, Northwestern and Riverside (ACHMNR) has observed Λ_c^+ baryons produced in diffractive-like reactions of the type pp → Xp where X contains a Λ_c^+. They trigger on a solitary high momentum proton in one hemisphere and six or more charged tracks in the other hemisphere. They then combined an identified K^- and proton with a π^+ meson and observed a 5σ peak in the plot of effective masses at 2260 MeV, with a width of 20 MeV, consistent with their experimental resolution. The narrow width and the absence of a peak in the $K^-p\pi^-$ channel rule out the possibility that this peak corresponds to the strong decay of an "ordinary" S = -1 baryon. They also see a 2σ signal at the same mass in the $\Lambda^0\pi^+\pi^+\pi^-$ channel. A cross section can be extracted since a branching ratio of 2.2 ± 1% for $\Lambda_c^+ \to K^-p\pi^+$ has recently been reported[18]. Using this, a differential cross section for Λ_c^+ production of $d\sigma/dx \sim 60$ μb (50% errors) can be inferred.

C. ACCDHW $\Lambda_c^+ \to K^*(890)p$ and $\Delta^{++}K^-$ (ref. 20)

In this experiment the Annecy, CERN, College-de-France, Dortmund, Heidelberg, Warsaw (ACCDHW) collaboration used the SFM triggered by a K^- with $P_T > 0.5$ GeV/c. A signal for $\Lambda_c^+ \to K^-p\pi^+$ was seen when the following event selections were made:

1) $M(K^-\pi^+) = 890 \pm 40$ MeV (i.e., K*(890))
2) $n_{obs} < 11$.
3) A leading system was required in the opposite hemisphere.
4) $X_{K^-} > 0.3$.
5) P_T of K^- balanced (to 0.2 GeV/c) in the same hemisphere.
6) P_T of $(K^-p\pi^+)$ system is $\gtrsim 1$ GeV/c.

They also observe signals in the Δ^{++} channel and see a signal in the $K^-p\pi^-$ channel when the SFM is triggered on a directly produced electron (no signal is seen when triggered on a positron as expected for diffractive $\Lambda_c^{+}\bar{D}^{-}$ production). Using branching ratios for $\Lambda_c^+ \to K^*p$ with $K^* \to K^-\pi^+$ of .25% and $\Lambda_c^+ \to \Delta^{++}K^-$ of 0.3% (from the Mark II results[18]), they infer differential cross sections (for x > 0.3) of $d\sigma/dx = 600$ μb from the K*p results and 550 μb from the $\Delta^{++}K^-$ results. These results have at least 50% errors but are still considerably higher than the 60 μb result of the ACHMNR collaboration.

D. UCLA-Saclay $\Lambda_c^+ \to K^-p\pi^+$ and $\Lambda^0\pi^+\pi^+\pi^-$

This group triggers on the production of at least 3 charged particles within 5% of one of the proton beams. Effective mass combinations for identified $K^-p\pi^+$ systems show a peak at a mass corresponding to the Λ_c^+. No peak is seen in the $K^- p\pi^-$ system consistent with the charm interpretation. A signal at the same mass is also seen in the $\Lambda^o\pi^+\pi^+\pi^-$ system. Using the reported branching ratio of 2.2 ± 1% [18] gives a differential cross section of $\frac{d\sigma}{dx}$ (0.75 ⩽ x ⩽ .9) = 620 μb. The errors here are again approximately 50%. These results are consistent with the results of the ACCDHW group and higher than those reported by the ACHMNR group.

Figure 6 shows the results of the three Λ_c^+ measurements compared with the cross section for the Λ production.[11] Recall that a naive guess for diffractive Λ_c^+ production was that it would be ∿1/10 that for Λ^o. The data are not very consistent but are perhaps not so inconsistent with this naive assumption.

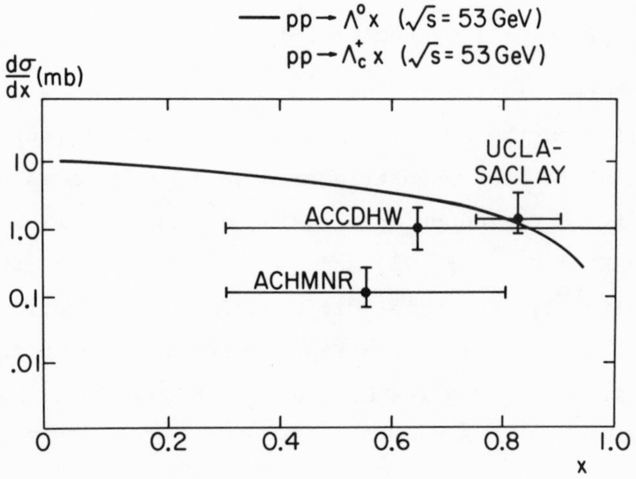

FIG. 6: The recent Λ_c^+ results from references 19, 20, and 21 compared with high energy inclusive Λ production data.

V. NEUTRINO BEAM DUMP EXPERIMENT, RECENT RESULTS

In the latest CERN beam dump experiment a multiple density target was used to allow an extrapolation to 0 decay length for inferring prompt neutrino yields. The incident proton beam was carefully instrumented to eliminate upstream scraping. Three detectors were in operation:

1) BEBC,[22] a large bubble chamber with a fiducial mass of 13 tons of Neon/Hydrogen. This detector could identify both muon and electron final states individually. Identification of electron events is particularly advantageous since prompt electron neutrinos do not have a severe background from π and K decays.

2) CERN, Dortmond, Heidelberg, Saclay (CDHS)[23]; A large iron detector with a fiducial mass of 500 tons. This detector could identify muon events and inferred electron events from a excess of "neutral current" events.

3) CERN, Hamburg, Amsterdam, Rome, Moscow (CHARM)[24]; A detector with a fiducial mass of 100 tons of marble plates. Again this detector identifies mu events and infers electron events from an excess of "neutral current" events.

In the beam dump run 0.3×10^{18} protons were used with the target density expanded ($\rho_{eff} \sim 1/3\ \rho_{cu}$) and 0.8×10^{18} protons with $\rho_{eff} = 1\ \rho_{cu}$. The signals extracted from the density extrapolation are in good agreement with the results using the density $= 1\ \rho_{cu}$ and using a computed background subtraction.

Results of μ^- events from CDHS are shown in Fig. 7. Also in the figure are the results from a Monte Carlo calculation which uses as an input a $D\bar{D}$ production cross section $\frac{Ed^3\sigma}{dp^3} \propto (1-x)^3 e^{-2P_T}$. The agreement is quite good.

The inferred cross sections for $D\bar{D}$ production from the three detectors are listed in Table I.

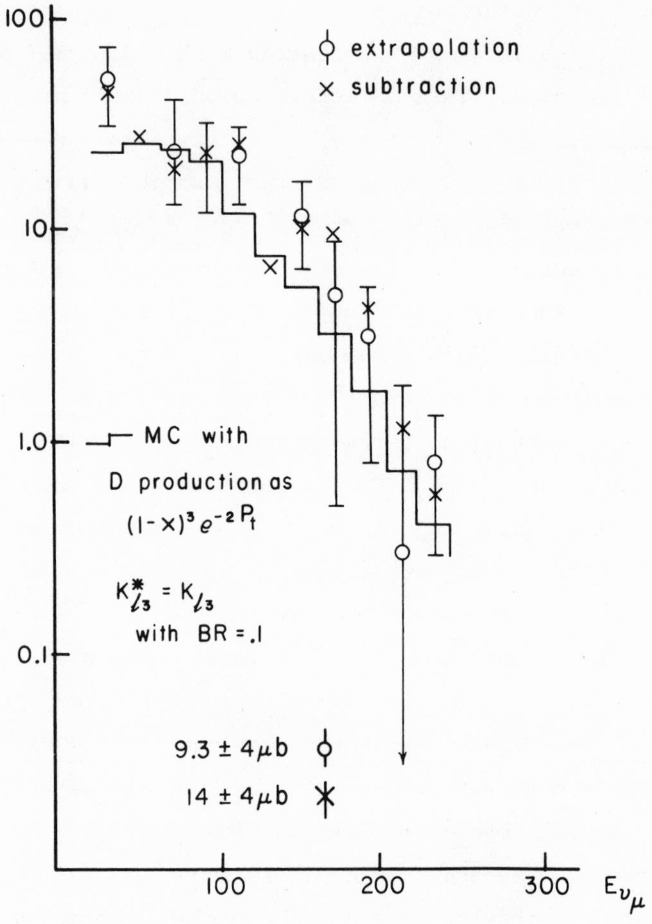

FIG. 7: Energy distribution of prompt ν_μ events from ref. 23 (CDHS).

TABLE I

$\sigma_{D\bar{D}}$ determination using $E \frac{d^3\sigma}{dp^3} \propto (1-x)^3 e^{-2P_T}$ as a production model,

and BR $D \rightarrow K^*\ell\nu$ 5%

$D \rightarrow K \ell\nu$ 5%

	$\sigma_{D\bar{D}}$ (μb)		
	BEBC	CDHS	CHARM
electron events	11 ± 2.1	-	14.5 ± 5.1
muon events	22 ± 6.4	9.3 ± 4 (ρ extrapolation)	29 ± 9
		14 ± 3 (computed subtraction)	23 ± 5

These results are consistent with the cross section of 14 μb reported earlier by the Cal Tech Stanford group[16] based on measurements of direct muon production with associated missing energy.

VI. DIRECT MUON PRODUCTION - RECENT RESULTS

A collaboration of Cal-Tech, Fermilab, Rochester and Stanford (CFRS)[25] has recently reported new results on direct muon production at Fermilab. In this experiment direct muon production was measured for all P_T and all positive values of x, in an experiment designed to accept 40% of the muons produced by semileptonic charmed meson decay. This large acceptance is needed in order to make charm production cross section determinations which are insensitive to the character of the differential cross section.

In this experiment a high precision calorimeter target precedes an array of large (3m × 3m) steel arrays interspersed with spark chambers and liquid scintillation counters (see Fig. 8). The detector was triggered on any proton interaction in the calorimeter which produced a muon which penetrated 5.8 m of iron, corresponding

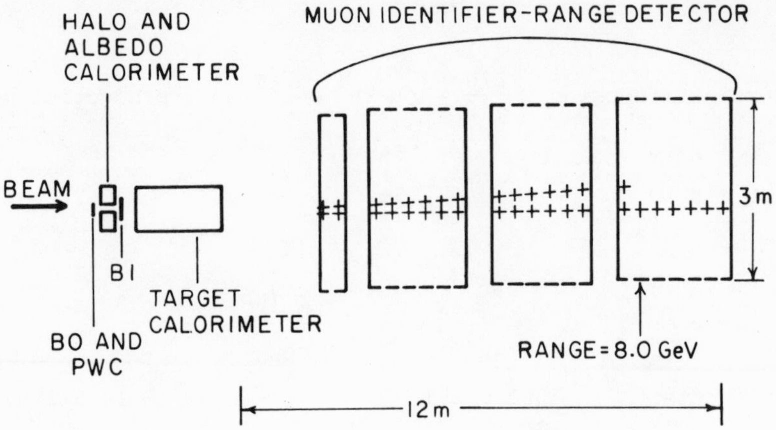

FIG. 8: Apparatus for the direct muon production experiment of Ritchie et al (ref. 25).

to a minimum energy of 8 GeV. The target calorimeter consisted of movable steel plates which enabled the variation of average target density from 1/4 to 3/4 the density of iron. The rate of directly produced muons could then be determined by extrapolating to infinite density (0 decay length). This extrapolation is shown in Fig. 9. Also shown in Fig. 9 is the rate, as a function of density, of dimuon events where both muons satisfy the trigger condition. These correspond to 36% of the direct muon events. The contribution to the remaining single muon rate from dimuons where the second muon was not seen was inferred by both looking for events where the second muon stopped before the 8 GeV trigger cut-off in the detector and extrapolating to 0 energy and by Monte Carlo calculations using known results on dimuon production. In this way a prompt single muon rate of 1.1×10^{-4} per p Fe interaction was inferred. Using a 40% acceptance and an 8% semileptonic branching ratio and assuming $\sigma_{D\bar{D}} \propto A^1$ gives a cross section for charm production

$$\sigma(\text{charm}) = 22 \pm 9\ \mu\text{b/nucleon}\ ,$$

which is rather insensitive to the particular production model.

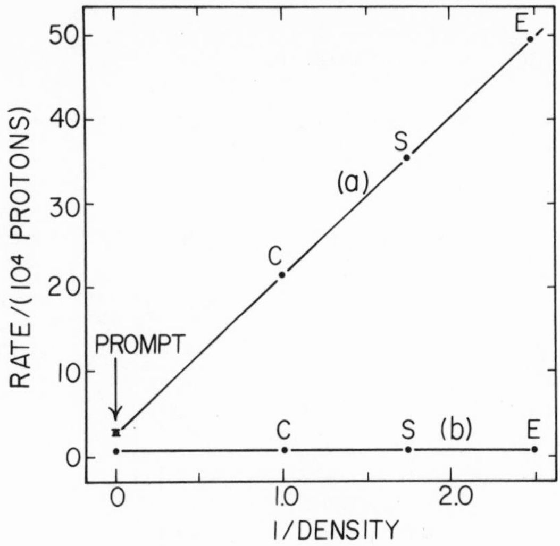

FIG. 9: The direct muon rate vs. density from ref. 25, showing the extrapolation to 0 decay length for a) single muon triggers and b) dimuon triggers.

VII. SUMMARY OF MESON RESULTS

A summary of charmed meson production results is contained in Table II. Included in this table are some earlier results not mentioned in this report.

As is apparent in Table II, the results from accelerator experiments are in reasonably good agreement, consistent with central charm production with a total charm production cross section of between 10 and 20 μb. The large "diffractive-like" cross sections for Λ_c^+ production at the ISR are not apparent in these lower energy D meson production results. The two ISR experiments addressing D meson production, the CCHK results[17] on $D^+ \rightarrow K^-\pi^+\pi^+$ and the earlier CERN, ETH, Saclay results[27] from μe coincidences are in strong disagreement.

TABLE II

(Assuming direct μ's and ν's originate from semileptonic D meson decays.)

Group	$\sqrt{s}$ (GeV)	<x>	$d\sigma/dx$ (μb)	$\sigma_{D\overline{D}}$ (μb)
CCHK[17]	53	> 0.3	55±22	∿ 110 $d\sigma/dx$∿flat
				∿1000 $d\sigma/dx$∿$(1-x)^3$
BEBC[22]	27	0	∿22	10-20
CDHS[23]	27	0	∿20	9-15
CHARM[24]	27	0	∿30	15-30
Cal-Tech[16] Stanford	27	0	∿30	14-20
CFRS[25]	25	0	∿40	22±9
CERN, ETH[26] Saclay	53	0	∿40	22±5 (e coincidence)
BEBC[27] (Track sensitive Tgt)	11.5 ($\pi^- p$)		5 direct electrons	23±13
Serpukov[28]	11.5 (pFe)		ν beam dump	5±4
Michigan[29]	27 (pW)		" " " (test run)	30-15

VIII. CONCLUSIONS

It is clear that charmed particles have been seen both at the ISR and at Fermilab and the SPS. The cross section at Fermilab and the SPS is around 10 to 20 μb, above the estimates from QCD. Large diffractive cross sections reported at the ISR have not been seen in accelerator experiments, although the accelerator energies are more than adequate to produce charmed states diffractively.

The cross sections reported at Fermilab/SPS correspond to approximately one charmed production event per 1000 proton interactions. Thus hadronic production experiments, if properly triggered, could prove to be valuable means for detailed studies of the properties of charmed particles. Also the semileptonic decays of hadronically produced charmed mesons would be a reasonably copious source for electron neutrinos.

IX. SOME COMMENTS ON THE HADRONIC PRODUCTION OF B MESONS

The Cal-Tech, Stanford collaboration[30] has recently reported some results on the production of B mesons in hadronic reactions. QCD estimates made by Halzen predict a cross section of $\sim$10 nb/nucleon for 400 GeV pp collisions. A crude guess that $\sigma_{B\bar{B}}/\sigma_{T} = \sigma_{D\bar{D}}/\sigma_{\psi} = 100$ predicts a 2 nb cross section for 400 GeV pp collisions, in accord with the QCD result. The Gluon Vector Dominance model of Fritzsch and Streng[9] predicts a rather large $\sigma_{B\bar{B}}$, $\sim$850 nb for 400GeV pp and 200 nb for 150 GeV πp collisions. (A B signal at the 200 nb level may have been seen by the CERN Goliath experiment[31] with a 150 GeV π beam.) Figure 10 indicates possible multi-muon signals from $B\bar{B}$ production and subsequent decays. Theory or "best guess" for branching ratios have been made where needed. Three possible multi-muon signals for $B\bar{B}$ production emerge:

1) $\psi + \mu$ with $\Psi \to \mu^+\mu^-$. Models have suggested that $B \to \psi\, x$ may have a branching ratio of as much as 3%.[32] Indeed it is in the Ψ channel that ref. 27 reports a result. An accompanying muon from the semileptonic decay of the other produced B particle or from the semileptonic decay of a daughter D meson is expected $\sim$10 to 20% of the time. The accompanying muon will typically have a $p_T \gtrsim 1$ GeV.

2) Same sign dimuons $\mu^{\pm}\mu^{\pm}$. These events arising from the semileptonic decay of one B particle and the semileptonic decay of a daughter D meson from the other B particle should be characterized by missing energy in the final state since two neutrinos are produced.

3) Three muons $\mu^+\mu^-\mu^{\pm}$. These events due to three semileptonic

decays of B's and daughter D's should have three neutrinos in the final state and thus considerable missing energy. The results from ref. 30 using the assumed branching ratios indicated in Fig. 10 are given in Table III. (The expected backgrounds come from measured $\pi \to \mu$ decays.)

Thus $\sigma_{B\bar{B}}$ in 400 GeV pp collisions is probably less than 50 nb, in contradiction to the Vector Gluon Dominance model. Similar measurements with pion beams are needed to confirm or dispute the Goliath results.

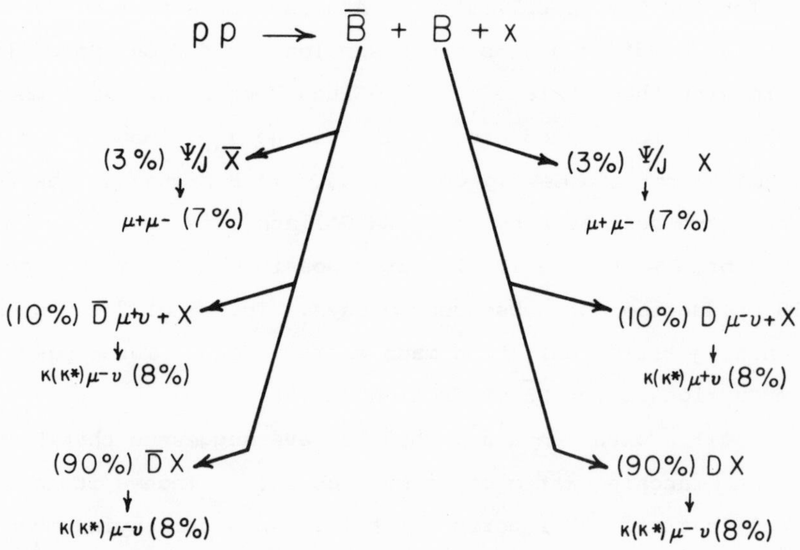

FIG. 10: A schematic representation showing various possible muon signatures from $B\bar{B}$ production and subsequent decays.

In any event experiments looking for multimuon signature should be suitable to push the limits on $B\bar{B}$ production down to the QCD predictions.

TABLE III

Results for multimuon signals for hadronic $B\bar{B}$ production. The decay branching ratios use are:

$BR(B \to \Psi + X)$	3%	(ref. 32)
$\Psi \to \mu^+\mu^-$	7%	
$BR(B \to D\mu\nu + X)$	10%	
$BR(B \to D + X)$	90%	
$BR(D \to K(K^*)\ )$	8%	(ref. 14)

Final State	# of events	Background	$\sigma_{B\bar{B}}$ (nb)	90% confidence limit
$\Psi\ \mu$	2	3±.5	-11±16	< 21 nb
$\Psi\ \mu$ ($P_{T_\mu} > 1.4$)	1	.06±.04	29±29	< 87 nb
$\mu^+\mu^-\mu^\pm$	19	17±3	15±40	< 96 nb
$\mu^+\mu^-\mu^\pm$ (missing energy > 30 GeV)	1	.7±.4	5±17	< 39 nb
$\mu^+\mu^+$	92	70±17	41±38	<116 nb
$\mu^+\mu^+$ (missing energy > 30 GeV)	8	6±2	8±15	< 38 nb

I would like to acknowledge the enthusiastic assistance of my colleague, A. Bodek, in the preparation of this review.

REFERENCES

1. J.J. Aubert et al., Phys. Rev. Lett. 33, 1414 (1974).
2. J.E. Augustin et al., Phys. Rev. Lett. 33, 1406 (1974).
3. G. Goldhaber et al., Phys. Rev. Lett. 37, 255 (1976).
4. E.G. Cazzoli et al., Phys. Rev. Lett. 34, 1125 (1975).

5. A. Benvenuti et al., Phys. Rev. Lett. 34, 419 (1975).

6. Fermilab Experimental Proposal #625, W. Lee, spokesman.

7. P.A. Rapidis et al., Phys. Rev. Lett. 39, 526 (1977).

8. Fermilab Experimental Proposal #515, J. Rosen, Spokesman.

9. F. Halzen, Proceedings of the Cosmic Ray and Particle Physics 1978 (Bartol Conference), Delaware, P. 261, T.K. Gaisser, editor.

10. This comes from integrating the fit to the dimuon cross section given in J.K. Yoh et al., Phys. Rev. Lett. 41, 684 (1978).

11. F. W. Büsser et al., Phys. Lett. 61B, 309 (1976); H. Kichini et al., Phys. Lett. 72B, 411 (1978).

12. H. Fritzsch and K. H. Streng, Phys. Lett. 78B, 447 (1978).

13. This topic was recently reviewed by A. Keenan at the Fermilab Lepton Photon Conference (August 1979).

14. W. Bacino et al., SLAC-Pub-2353, June 1979.

15. This topic was recently reviewed by H. Wachsmuth at the Fermilab Lepton Photon Conference (August 1979).

16. K. W. Brown et al., Phys. Rev. Lett. 34, 410 (1979); A.M. Diament-Berger et al., Phys. Rev. Lett. 43, 1773 (1979).

17. D. Drijard et al., Phys. Lett. 81B, 250 (1979).

18. G.S. Abrams et al., Phys. Rev. Lett. 44, 10 (1980). See also J. Kirby, Report at the Fermilab Lepton Photon Conference (August 1979).

19. K.L. Giboni et al., Phys. Lett. 85B, 437 (1979).

20. D. Drijard et al., Phys. Lett. 85B, 452 (1979).

21. W. Lockman et al., Phys. Lett. 85B, 443 (1979).

22. P. Renton, Proc. Neutrino 79, Bergen, Norway, June 1979.

23. V. Hepp, Proc. Neutrino 79, Bergen, Norway, June 1979.

24. W. Kozanecki, Proc. Neutrino 79, Bergen, Norway, June 1979.

25. J. Ritchie et al., Phys. Rev. Lett. 44, 230 (1980).

26. A. Chilingarov, et al., Phys. Lett. 83B, 136 (1979).

27. R. Barloutod et al., submitted to the Fermilab Lepton Photon Conference, August 1979.

28. Asratyan et al., Phys. Lett. 79B, 497 (1978).

29. B.P. Roe et al., Univ. of Michigan report UMHE 79-2.

30. A.M. Diament-Berger et al., "Search for Possible Signatures of Bottom-Quark States Produced in 400 GeV p-Fe Interactions."

31. R. Barati et al., CERN EP/79-113, submitted to the International Symposium on Lepton and Photon Interactions at High Energy, Batavia, IL. August 1979.

32. H. Fritzsch, CERN Preprints, TH2648 (March 1979) and TH2703 (July 1979).

INFRARED PROPERTIES OF THE GLUON PROPAGATOR: A PROGRESS REPORT

F. Zachariasen

California Institute of Technology

Pasadena, California 91125

In several earlier publications[1], we have described a program for investigating the infrared behavior of non-Abelian gauge theories. Here we wish to report further progress toward this goal.

The approach is based on the Dyson equation for the gluon propagator in axial gauge, coupled with the Ward identities to determine the longitudinal parts of the vertex functions appearing in the Dyson equation. The unknown transverse parts of the vertex functions are assumed to be irrelevant in the IR regime, as is known to be the case in QED and in one loop QCD. We ignore quarks and work with a pure glue theory.

To simplify the spinology, we make an (in principle and perhaps in practice verifiable) ansatz, to the effect that the spin structure of the IR singular part of the propagator is the same as that of the free propagator. (This is also true in QED.)

The above assumption and ansatz, with no further restrictions, lead to an integral equation for the gluon propagator:

$$q^2 \frac{(1-1/\gamma)}{Z(q)} = q^2(1-1\gamma) + g_0^2 \int dk\, K_{\sigma\sigma'}(k,k',n) \cdot$$

$$\{- \frac{Z(k)}{Z(q)} \frac{Z(k')-Z(q)}{k'^2-q^2} (q+k')_\sigma q_{\sigma'} \tag{1}$$

$$+ \frac{Z(k)-Z(k')}{k^2-k'^2} (k\cdot k' \delta_{\sigma\sigma'} - k'_\sigma k_{\sigma'}) + Z(k)\delta_{\sigma\sigma'}\} + C(\gamma) \quad .$$

In this equation, $Z(q)$ is defined in terms of the gluon propagator $\Delta_{\mu\nu}(q)$ by

$$\Delta_{\mu\nu}(q) = - \frac{Z(q)}{q^2} (\delta_{\mu\nu} - \frac{q_\mu n_\nu + n_\mu q_\nu}{n\cdot q} + \frac{q_\mu q_\nu n^2}{(n\cdot q)^2})$$

$$= Z(q)\Delta^{(0)}_{\mu\nu}(q) \quad , \tag{2}$$

(we suppress color indices), n_μ is the direction of the gauge choice, and $\gamma \equiv n^2q^2/(n\cdot q)^2$. The kernel $K_{\sigma\sigma'}$ is given by

$$K_{\sigma\sigma'}(k,k',n) = \frac{n\ (k-k')n\cdot k'}{n^2} \Delta^{(0)}_{\lambda\sigma}(k)\Delta^{(0)}_{\lambda\sigma'}(k') \quad , \tag{3}$$

where $k+k' = q$. Finally, from gauge invariance one can show that $C(\gamma)$ is such that the right-hand side of the integral equation vanishes as the four vector $q_\mu \to 0$.

Our hope is that the solution to this integral equation has the same IR behavior as the true gauge theory propagator.

The integrals in (1) have ultraviolet divergences which we limit with a cutoff Λ. This provides a mass scale in the problem, so that the solution is a function of two independent dimensionless scalars: $Z(q) = Z(q^2/\Lambda^2,\gamma)$. As we will see shortly, after renormalization, the renormalized Z will be a function of a finite scale M: $Z_R(q) = Z_R(q^2/M^2,\gamma)$. We expect that the UV and IR limits of Z_R

will be independent of γ and thus gauge invariant.

To begin to study the properties of (1), suppose we insert on the righthand side $Z(q) = (A/q^2)^\alpha$ for some $\alpha > 0$. At first glance it might be thought that we can only do this for $\alpha < 1$, since otherwise divergences like $\int d^4k/k^4$ for small k^2 will appear. In fact these are absent, basically because the angular average of $\Delta^{(0)}_{\mu\nu}(k)$ is zero. Thus we can allow $\alpha \leq 1$.

With this input, the RHS of (1) can be expended for small q^2, and one finds a series of the form

$$q^2(1-1/\gamma) + C_0(\gamma,\alpha)q^2 \left(\frac{A}{q^2}\right)^\alpha + C_1(\gamma,\alpha)q^2 \left(\frac{A}{q^2}\right)^{\alpha-1} + \ldots + C(\gamma) \quad . \qquad (4)$$

For consistency with the LHS this must be of order $(q^2)^{\alpha+1}$. The first requirement, therefore, is that $C_0(\gamma,\alpha) = 0$ for all γ. The next requirement is that the free inverse propagator term $q^2(1-1/\gamma)$ must be cancelled. This suggests (though it does not prove) that α is an integer. So let us try $\alpha = 1$. Then the C_0 term is just $C_0(\gamma,\alpha)\cdot A$, a constant as $q^2 \to 0$. But it must, by gauge invariance, then be cancelled by $C(\gamma)$. Thus the constraint $C_0(\gamma,\alpha) = 0$ becomes, for $\alpha = 1$, $C_0(\gamma,1) + C(\gamma) = 0$, a statement which is already guaranteed by gauge invariance.

The next requirement now reads $(1-1/\gamma) + C_1(\gamma,1) = 0$. After that the $O(q^4)$ term on the RHS must match the $O((q^2)^{\alpha+1}) = O(q^4)$ term on the LHS, and so forth.

The conclusion of this discussion is that $\alpha = 1$(i.e., $Z(q) \to A/q^4$ as $q^2 \to 0$) is an attractive possibility. We note, for later reference, that if we put in $Z(q) \equiv A/q^2$ for all q^2 on the RHS, then the integrals yield just function of γ independent of q^2 which will be cancelled by $C(\gamma)$. There are no logarithms with this input, again because of the fact that the angular average of $\Delta^{(0)}_{\mu\nu}(k)$ is zero. This fact is crucial for the existence of a solution behaving like $1/q^2$ as $q^2 \to 0$. We note that without the ansatz (2), this angular average would not vanish, and one would in general expect logarithms to result from a $1/q^2$ input. Furthermore, an input with $\alpha = 0$ will also produce logarithms.

To proceed to explore whether or not a solution approaching $1/q^2$ in the IR actually exists it is appropriate, at this point, to renormalize the integral equation. We may do this as follows. Let us rewrite eq. (1) in the form[2]

$$\frac{1}{Z(q)} = 1 + g_0^2 \int K(k,q)Z(k)dk + \frac{g_0^2}{Z(q)} \int L(k,q)Z(k)Z(q-k)dk \quad . \tag{5}$$

We imagine all "quadratic divergences," that is, contributions to the RHS of (1) behaving like constants as $q \to 0$ to have been already isolated in (5) and cancelled against $C(\gamma)$. Thus the worst divergences in (5) are logarithmic. We next define $Z(q) = Z(M)Z_R(q)$ for some fixed (spacelike) four-vector M_μ, and normalize $Z_R(M) = 1$. The renormalized coupling constant is defined by

$$g^2(M) = \frac{g_0^2\, Z(M)}{1+g_0^2 Z(n) \int dk\; K(k,k',n)Z_R(k)} \quad , \tag{6a}$$

which, in virtue of (5), can also be written

$$g^2(M) = \frac{g_0^2 Z^2(M)}{1-g_0^2 Z^2(M) \int dk\; L(k,k',M)Z_R(k)Z_R(k')} \quad . \tag{6b}$$

Then it is easy to show that

$$\frac{1}{Z_R(q)} = 1+g^2(M) \int [K(k,q)-K(k,M)]Z_R(k)\; dk + \frac{g^2(M)}{Z_R(q)} \int [L(k,q)Z_R(q-k)-L(k,M)Z_R(M-k)]Z_R(k)\; dk \quad . \tag{7}$$

In (7), all integrals are now finite.

In perturbation theory, we find from (7) that

$$Z_R(q) = 1 - g^2(M)b \log q^2/M^2 \tag{8}$$

and we find from (6) that

$$g^2(q) = g^2(M) - g^4(M)(2b-c)\log q^2/M^2 \ , \tag{9}$$

where

$$\int [K(k,q) - K(k,M)]dk = (b-c)\log q^2/M^2 \ ,$$

and

$$\int [L(k,q) - L(k,M)]dk = c \log q^2/M^2 \quad .$$

Therefore we can anticipate the exact asymptotic behavior as $q^2/M^2 \to \infty$ to be

$$g^2(q) \to \frac{g^2(M)}{1+g^2(M)(2b-c)\log q^2/M^2} \quad , \tag{10}$$

and

$$Z_R(q) \to \left(\frac{1}{1+g^2(M)(2b-c)\log q^2/M^2}\right)^{\frac{b}{2b-c}} \quad . \tag{11}$$

Explicit calculation gives $b/c = 11/6$ (b is the usual b of non-Abelian gauge theories: $b = \frac{1}{16\pi^2} \cdot \frac{11}{3} \cdot C_A$). Thus we expect

$$Z_R(q) \to \left(\frac{1}{1+g^2(M)\ \frac{16}{11}\ b \log q^2/M^2}\right)^{11/16} \quad . \tag{12}$$

To illustrate those points, it is instructive to study a simple analytically soluble model of the ultraviolet behavior of the integral equation. As the model unrenormalized equation, choose

$$\frac{1}{Z(q)} = 1 - (b-c)g_0^2 \int_{q^2}^{\Lambda^2} \frac{dk^2}{k^2} Z(k) - \frac{c\, g_0}{Z(q)} \int_{q^2}^{\Lambda^2} \frac{dk}{k^2} Z^2(k) \quad . \tag{13}$$

In perturbation theory, $Z(q) = 1 + b\, g_0^2 \log q^2/\Lambda^2 + \ldots$. One can easily verify that

$$Z(q) = \left(1 + (2b-c)g_0^2 \log q^2/\Lambda^2\right)^{-\frac{b}{2b-c}} \tag{14}$$

solves this nonlinear integral equation exactly. The renormalized equation corresponding to (13) is

$$\frac{1}{Z_R(q)} = 1 + (b-c)g^2(M) \int_{M^2}^{q^2} \frac{dk^2}{k^2} Z_R(k) + \frac{c\, g^2(M)}{Z_R(q)} \int_{M^2}^{q^2} \frac{dk^2}{k^2} Z_R^{\,2}(k) \quad , \tag{15}$$

with the renormalized coupling constant defined by

$$g^2(M) = \frac{g_0^2\, Z^2(M)}{1 + c\, g_0^2\, Z^2(M) \int_{M^2}^{\Lambda^2} \frac{dk^2}{k^2} Z_R(k)} \quad , \tag{16}$$

or equivalently,

$$g^2(M) = \frac{g_0^2 \, Z(M)}{1-(b-c)g_0^2 Z(M) \int_{M^2}^{\Lambda^2} \frac{dk^2}{k^2} Z_R(k)} \quad . \tag{17}$$

The solution to the renormalized integral equation is, of course,

$$Z_R(q) = \left(1+(2b-c)g^2(M)\log q^2/M^2 \right)^{-\frac{b}{2b-c}} \quad ; \tag{18}$$

The renormalized coupling constant scales like

$$g^2(q) = \frac{g^2(M)}{1+(2b-c)g^2(M)\log q^2/M^2} \quad . \tag{19}$$

All the features of the actual equation are displayed explicitly by this model.

We are now in a position to rewrite (1) explicitly, as a finite equation, with a known UV asymptotic behavior, in a form suitable for numerical computations.

First we must remove the parts cancelled by $C(\gamma)$. There are two such - one from the UV (the "quadratically divergent" part) and one from the IR corresponding to the guess that $Z \rightarrow A/q^2$ in the IR. The UV part is easily calculated from (1) and is given by

$$\int dk \; K_{\sigma\sigma'}(k,-k,n) \left[Z(k)\delta_{\sigma\sigma'} + \frac{\partial Z(k)}{\partial k^2} (-k^2\delta_{\sigma\sigma'} + k_\sigma k_{\sigma'}) \right]$$

$$= -3 \int dk/k^2 \left[Z(k) + k^2 \frac{\partial Z(k)}{\partial k^2} \right] \quad . \tag{20}$$

(Note, incidentally, that if $Z(k) = A/k^2$, this vanishes.) The IR part is isolated by writing $Z(k) = Z_1(k) + A/k^2$, where Z_1 is finite in the IR, under the integral in (1), and then adding and subtracting the same integrand with Z replaced by A/k^2. The added term is cancelled by $C(\gamma)$, leaving no remainder; the subtracted integrand now generates no self mass. At this point, the renormalization procedure described above may be used to generate from (1) a finite equation with a finite mass scale.

Let us suppose this equation turns out (as we expect it to) to possess a solution in which $Z(q) \to A/q^2 \to 0$. What implications does this have for confinement? To explore this question, let us first calculate the coordinate space propagator at large distances. We write

$$\Delta_{\mu\nu}(x) = \int \frac{d^4q}{(2\pi)^4} e^{iqx} \Delta_{\mu\nu}(q) \tag{21}$$

and choose

$$\Delta_{\mu\nu}(q) = \frac{1}{q^4}\left(\delta_{\mu\nu} - \frac{q_\mu n_\nu + n_\mu q_\nu}{q\cdot n} + \frac{q_\mu q_\nu n^2}{(q\cdot n)^2}\right). \tag{22}$$

A straightforward calculation yields

$$\Delta_{\mu\nu}(x) = \frac{1}{8\pi^2}\left\{\frac{x_\parallel}{x}\tan^{-1}\frac{x_\parallel}{x}\left(\delta_{\mu\nu} - \frac{n_\mu n_\nu}{n^2} - \frac{(x_\perp)_\mu (x_\perp)_\nu}{x_\perp^2}\right) - \frac{(x_\perp)\,(x_\perp)}{x_\perp^2}\right\} , \tag{23}$$

where

$$x_\parallel = \frac{n\cdot x}{n} , \tag{24}$$

and

$$(x_{\shortparallel})_\mu = x_\mu - \frac{n\cdot x}{n^2} n_\mu \quad . \tag{25}$$

Note in spite of the fact that superficially the integral in (21) diverges at $k = 0$, there are no logarithms in $\Delta_{\mu\nu}(x)$. This is again because of the vanishing angular average of $\Delta_{\mu\nu}(k)$, which in fact makes (21) exist at $k = 0$ without cutoff.

We may also calculate the three-space propagator: we find, for large $(\vec{x})$,

$$\Delta_{oo}(x) \equiv \int \frac{d^3\vec{q}}{(2\pi)^2} e^{-i\vec{q}\cdot\vec{x}} \Delta_{oo}((0,\vec{q})) \sim |\vec{x}| \quad . \tag{26}$$

The first nontrivial term in the Wilson loop is related to $\Delta_{\mu\nu}(x)$:

$$\langle \mathrm{Tr}\, e^{\tau a \oint dx_\mu A_\mu^a(x)} \rangle = 1 + \mathrm{Tr}\, \tau^a \tau^b \int dx_\mu \oint dx'_\nu \Delta^{ab}_{\mu\nu}(x-x) + \ldots \quad . \tag{27}$$

We first note that (23) can be rewritten in the form

$$\Delta_{\mu\nu}(x) = \frac{1}{16\pi^2} \{ n\cdot\frac{\partial}{\partial x} \delta_{\mu\nu} - n_\nu \frac{\partial}{\partial x_\mu} - n_\mu \frac{\partial}{\partial x_\nu} - n^2 \frac{\partial}{\partial n_\mu} \frac{\partial}{\partial x_\nu} \} \cdot$$

$$\cdot \left[\frac{1}{n} \left(x_\shortparallel \log x^2 - 2x_\shortparallel + 2x_\perp \tan^{-1} \frac{x_\shortparallel}{x_\perp}\right)\right] \quad . \tag{28}$$

Hence (26) becomes, for large area,

$$\langle \mathrm{Tr}\, e^{\tau^a \oint dx_\mu A_\mu^a(x)} \rangle = 1 + \mathrm{Tr}(\tau^a \tau^a) \cdot \frac{1}{16\pi^2} \cdot \pi \cdot A + \ldots \quad , \tag{29}$$

where A is the loop area. As it should be, the result is independent of n. If higher order terms in the expansion behave in the analogous way, as they would for example if disconnected parts of the higher terms dominate for large A, then we have confinement.

Finally, we comment that a propagator behaving like $1/q^4$ also implies confinement if we take seriously the leading log perturbation theory results for near mass shell Green's functions obtained by Cornwall and Tiktopoulos[3].

All of these arguments, then, suggest that the result $Z(q) \to A/q^2$ as $q^2 \to 0$ indeed implies confinement.

REFERENCES

1. R. Anishetty et al., Physics Letters 86B (1979) 52;
   M. Baker et al., CALT-68-741, September 1979;
   F. Zachariasen, CERN TH-2601 (1978);
   M. Baker, Univ. of Washington RLO-1388-781 (1979);
   J.S. Ball and F. Zachariasen, Nucl. Phys. B143 (1978) 148.
2. A simplified version of this equation, in which the L term is dropped and the calculations are done in covariant gauge, has been explored by S. Mandelstam, UCB-PTH-79/8. He finds results similar to ours.
3. J.M. Cornwall and G. Tiktopoulos, Phys. Rev. D10 (1977) 2937.

THEORETICAL ASPECTS OF PROTON DECAY*

William J. Marciano

The Rockefeller University

New York, New York 10021

ABSTRACT

Theoretical predictions for the proton's lifetime, τ_p in grand unified theories are reviewed. A simple procedure for computing M_S, the super-heavy mass of the vector bosons which mediate proton decay in the Georgi-Glashow SU(5) model, is outlined. Using the current value of α_s the QCD coupling, obtained from electroproduction as input, the SU(5) model is found to predict $M_S=8.5\times10^{14}$ GeV, $\sin^2\theta_W=0.207$ and $\tau_p\simeq1\text{-}20\times10^{31}$ yr.

I. INTRODUCTION

Professor Goldhaber[1] has already presented in his talk a thorough review of both the historical background and present experimental status of searches for proton decay. I will primarily concentrate on reviewing theoretical aspects of proton decay in the Georgi-Glashow SU(5) model[2] and its generalizations. In so doing, I will take this opportunity to describe the results of a recently completed calculation[3] of M_S, the super-heavy mass of vector bosons which mediate proton decay.

*Work supported in part by the U.S. Department of Energy under Contract Grant No. EY-76-C-02-2232B.*000.

Let me begin by reminding you of the expectations of proposed searches for proton decay. The present bound on the proton lifetime is[4]

$$\tau_p \geq 10^{29} \sim 10^{30} \text{ yr.} \tag{1}$$

Recently proposed experiments[1,5] plan to search for proton decay up to about

$$\tau_p \simeq 10^{33} \text{ yr.} \tag{2}$$

At 10^{33} yr they can just manage to measure the lifetime; however, if τ_p is in the range $10^{30} \sim 10^{32}$ yr they will be able to determine branching ratios, final state polarizations, etc., and thereby rule out some of the competing grand unified models. Theorists can contribut to this effort by addressing questions such as: Do grand unified models predict $\tau_p \lesssim 10^{33}$ yr and if so, by how much? What are the dominant decay modes of the proton?

In section II of this talk I will briefly outline some features of the Georgi-Glashow SU(5) model and its generalizations. Predictions for τ_p as a function of M_S are reviewed and anticipated branching ratios of the various proton decay modes are discussed. Then in section III, I describe in some detail how M_S is obtained from calculations of radiative corrections in grand unified theories (GUTS). Employing a simple new procedure, I find $M_S \simeq 8.5 \times 10^{14}$ GeV for the current value of α_s, the QCD coupling, obtained from electroproduction experiments. This mass value implies $\tau_p \simeq 1\text{-}20 \times 10^{31}$ yr and predicts $\sin^2\theta_W \simeq 0.207$. Section IV summarizes and discusses the present theoretical status of proton decay calculations in the SU(5) model.

II. THE GEORGI-GLASHOW SU(5) MODEL

1. Basic Features. The SU(5) model of Georgi and Glashow[2] is the most economical of all grand unified theories (GUTS).[6] It is

minimal in that only the usual fermions are required; they make up three sequential generations, each of which is composed of a $\underline{5}+\underline{10}$ representation of SU(5). For example the lightest generation multiplet assignments are

$$\begin{pmatrix} d_1 \\ d_2 \\ d_3 \\ e^+ \\ \bar{\nu}_e \end{pmatrix}_R \qquad \sqrt{\frac{1}{2}} \begin{pmatrix} 0 & \bar{u}_3 & -\bar{u}_2 & -u_1 & -d_1 \\ -\bar{u}_3 & 0 & \bar{u}_1 & -u_2 & -d_2 \\ \bar{u}_2 & -\bar{u}_1 & 0 & -u_3 & -d_3 \\ u_1 & u_2 & u_3 & 0 & -e^+ \\ d_1 & d_2 & d_3 & e^+ & 0 \end{pmatrix}_L \tag{3}$$

Similarly, there is a second generation containing ν, ν_μ, s, c and a third composed of τ, ν_τ, b, t (with Cabibbo mixing among generations).

There are 24 gauge bosons in the SU(5) model. Twelve of these are the ordinary gauge fields (8 gluons, $W^\pm$, Z^0 and γ) while the other twelve are very exotic in that they carry color and have fractional electric charges $\pm$ 4/3 and $\pm$1/3. The latter belong to $SU(3)_c$ triplets and form an SU(2) doublet; hence they are degenerate up to very small SU(2) breaking $M_X \simeq M_Y = M_S$. These bosons and the pattern of symmetry breaking in the SU(5) model are illustrated in fig. 1.

SU(5)	24 Gauge Bosons
↓	(12 super-heavy, $X^{\pm 4/3}$ & $Y^{\pm 1/3}$) ($M_S \simeq 10^{15}$ GeV)
$SU(3)_c \times SU(2) \times U(1)$	($W^\pm$, Z^0; $M \simeq 10^2$ GeV)
↓	
$SU(3)_c \times U(1)$	(8 gluons, photon; all massless)

Fig. 1. Pattern of symmetry breaking and resulting vector boson mass scales in the Georgi-Glashow SU(5) model.

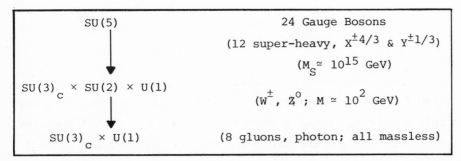

The minimum Higgs scheme required to break the gauge symmetry and provide fermion masses is a real 24-plet and a complex 5-plet (there may be many such 5-plets).[7] In addition, to get the fermion masses right, a 45-plet of Higgs scalars may be required.[2] The masses of the physical scalar particles in this model are somewhat arbitrary. I assume[8] that all physical scalars originating from the 24-plet and all fractionally charged scalars coming from 5-plets (and 45's) have super-heavy mass M_S, while the N_H iso-doublets under SU(2) weak isospin have mass $\simeq M_W$.

Some nice features of the SU(5) model are:[9] It provides a natural explanation for charge quantization, massless neutrinos and observed fermion mass ratios.[7,10] Furthermore, this theory is truly unified in that the bare couplings associated with the low energy subgroups $SU(3)_c \times SU(2) \times U(1)$ are constrained to be equal

$$g_{3_0} = g_{2_0} = g_{1_0} \quad . \tag{4}$$

This implies for the SU(5) model that the bare electric charge, e_0 is given by

$$\frac{1}{e_0^2} = \frac{5}{3g_{1_0}^2} + \frac{1}{g_{2_0}^2} = \frac{8}{3g_{2_0}^2} \tag{5}$$

and consequently that $\sin^2\theta_W^0$, the bare weak mixing angle, is given by

$$\sin^2\theta_W^0 = \frac{e_0^2}{g_{2_0}^2} = \frac{3}{8} \quad , \tag{6}$$

a rational number rather than an infinite adjustable counterterm; its role in the Weinberg-Salam model[11] when considered alone. The relationships between the bare quantities in (4)-(6) are natural (in the technical sense); hence similar relationships exist between the renormalized couplings up to finite calculable higher order corrections.[12] Therefore, defining $\alpha_i = g_i^2/4\pi$, $\alpha = e^2/4\pi$ and identifying $\alpha_3 = \alpha_s$, the QCD coupling, one finds

$$\frac{\alpha}{\alpha_s} = \frac{3}{8}[1 - 0(\frac{\alpha}{\pi} \ell n \frac{M_S}{\mu})] \quad , \tag{7}$$

$$\sin^2\theta_W = \frac{3}{8}[1 - 0(\frac{\alpha}{\pi} \ell n \frac{M_S}{\mu})] \; , \tag{8}$$

where the mass scale μ in the radiative corrections to (7) and (8) depends on particular definitions of the renormalized couplings and $\sin^2\theta_W$ employed.[13] The calculated logarithmic corrections in (7) and (8) when inverted can be used to obtain M_S. (More about this procedure in section III.)

An important feature of the SU(5) model and many of the more general GUTS is baryon and lepton number nonconservation (although in the SU(5) model B-L is conserved). When this property is combined with hard CP violoation, it offers a nice explanation for the observed matter-antimatter asymmetry of the universe within the framework of big-bang cosmology.[14] That is, GUTS provide a viable explanation for the observed ratio

$$\frac{N_B}{N_\gamma} = 10^{-8} - 10^{-9} \quad , \tag{9}$$

where N_B is the excess of baryons and N_γ is the background radiation of the universe. This success provides the best available evidence for baryon number nonconservation. It implies that the proton should have a finite lifetime. (Unless accidentally stabilized by some unknown dynamical mechanism or peculiar fermion mixing pattern.)

2. Proton Lifetime. The exotic super-heavy gauge bosons $X^{\pm 4/3}$ and $Y^{\pm 1/3}$ in fig. 1 mediate baryon number violating processes such as proton decay. Some of the proton decay modes in the SU(5) model are schematically illustrated in fig. 2.

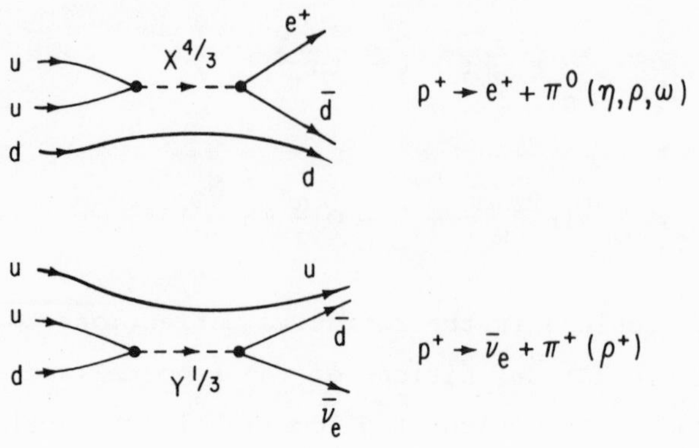

Fig. 2. Some proton decay channels in the SU(5) model.

General predictions regarding proton decay that transcend the SU(5) model follow merely from the assumption that $SU(3)_c \times SU(2) \times U(1)$ is respected by proton decay amplitudes.[15,16] For example, B-L is conserved up to possible violations of relative order $\alpha m_L/m_H$ where m_L/m_H is a light to heavy mass ratio. Furthermore, this symmetry requires that the vector boson mediators of proton decay belong to $SU(3)_c$ triplets and SU(2) doublets; there are two possible charge assignments for these allowed doublets

$$(X^{4/3},\ Y^{1/3}) \text{ and } (X'^{2/3},\ Y'^{-1/3}) \tag{10}$$

(There can potentially be several such doublets.) Only the first doublet exists in the SU(5) model while both are present in the SU(10) model. Similarly, scalar mediators must form SU(3) triplets and SU(2) triplets with charge assignments

$$(\phi^{2/3},\ \phi^{-1/3},\ \phi^{-4/3}) \quad . \tag{11}$$

Present day calculations in the SU(5) model (to be subsequently described) estimate $M_S \simeq M_X \simeq M_Y$ and the rate of proton decay due to those mediators. (Scalars are assumed to be unimportant.) Such estimates carry over to larger gauge groups which contain the SU(5) model (such as SO(10)), if $M_{X,Y} << M_{X',Y'}$. If $M_{X'}$ and $M_{Y'}$ are lighter than M_S, then τ_p will be even shorter than predicted by the SU(5) model.

There is a variety of ingredients which go into the theoretical prediction for the proton lifetime.[17] Included are estimates of the wavefunction overlap of two quarks in the proton, matrix elements of baryon number violating amplitudes and enhancement factors due to box diagrams involving the exchange of a super-heavy boson and one or more ordinary light vector bosons (gluons, γ, $W^{\pm}$, Z^{o}). The last of these effects has been calculated by Buras, Ellis, Gaillard and Nanopoulos;[15,18] for typical values of low energy couplings and three generations of fermions it enhances the amplitudes by about a factor of 3.7. In addition, the amplitude is proportional to $g^2(M_S)/M_S^2$ where $g(M_S)$ is the value of the effective SU(5) coupling at M_S; $g^2(M_S)/4\pi \simeq 0.024$. Including all these effects, the proton lifetime as a function of M_S has been calculated by several groups.[19,20,21] At present τ_p is estimated to be

$$\tau_p \simeq 2 \times 10^{-29} (M_S \text{ in GeV})^4 \text{ yr} \quad \left\{ \begin{matrix} \text{Jarlskog \& Yndurain}^{19} \\ \text{Din, Girardi \& Sorba}^{20} \end{matrix} \right\} , \qquad (12)$$

$$\tau_p \simeq 38 \times 10^{-29} (M_S \text{ in GeV})^4 \text{ yr} \quad \{\text{Donoghue}^{21}\} \qquad (13)$$

These calculations differ by a factor of 19; presumably this discrepancy will be at least partially eliminated by ongoing reanalysis and careful checking of the calculations. I take the point of view that the final SU(5) prediction will lie between the results of (12) and (13); therefore, I will generally quote a range of values for τ_p corresponding to these distinct results when I subsequently estimate the proton lifetime.

At this time there also exists some discrepancy concerning the SU(5) predicted branching ratios for the various proton decay modes The results of three different sets of calculations are illustrated in Table I.

Table I. Predicted branching ratios for the (strangeness conserving) two body decay modes of the proton in the SU(5) model obtained by three different calculations.

Decay Mode	Machacek[15] SU(6)	Donoghue[21] Bag Model	Din et al.[20] Bag Model
$p \to \pi^0 + e^+$	34.6%	9%	15 - 31%
$p \to \rho^0 + e^+$	17.3%	21%	32 - 21%
$p \to \eta + e^+$	12.1%	3%	4 - 5%
$p \to \omega + e^+$	22.5%	56%	29 - 19%
$p \to \pi^+ + \bar{\nu}$	8.9%	3%	5 - 11%
$p \to \rho^+ + \bar{\nu}$	4.6%	8%	12 - 8%

Donoghue's analysis clearly exhibits a suppression of the pion decay mode; an unfortunate finding, since that is one of the primary decay channels being experimentally looked for.[1] The two sets of predictions attributed to Din et al. correspond to their results without -- with phase space effects included; an indication of the degree of uncertainty. Pion suppression is not clear from their analysis. Is the pion decay mode actually suppressed? This important issue needs to be resolved.

III. THE SUPER-HEAVY MASS M_S

The final ingredient necessary to estimate τ_p from (12) or (13) is the value of the super-heavy mass M_S. That quantity is extracted in the SU(5) model from calculations of higher order radiative corrections to the natural relationships in (7) and (8); an approach initiated by Georgi, Quinn and Weinberg.[12] M_S has been estimated in two distinct ways. 1) By comparing the experimental value of α/α_S with theory as in (7). 2) By comparing the experimental value of $\sin^2\theta_W$ with theory as in (8). The first approach has been diligently pursued by Goldman and Ross; their latest estimate is[22]

$$M_S \simeq 4.4 \times 10^{14} \text{ GeV} \quad . \tag{14}$$

In the second approach, one uses the present world average[23]

$$\sin^2\theta_W = 0.23 \pm 0.015 \qquad \text{(Experiment)} \tag{15}$$

to infer in the SU(5) model[8]

$$0.20 \leq \sin^2\theta_W \leq 0.21 \quad \rightarrow \quad 10^{33}\text{ yr} \geq \tau_p \geq 10^{30}\text{ yr} \tag{16}$$

(The predicted lifetime gets too short if $\sin^2\theta_W > .21$.) I must however caution you that the experimental results in (15) may be slightly modified when radiative corrections of order α are fully accounted for (see ref. 13).

I have recently reexamined the radiative corrections in (7) and (8) employing a new very simple approach[3] which I will now briefly describe. To begin with I observe the bare couplings in the SU(5) model are equal

$$\alpha_{1_0} = \alpha_{2_0} = \alpha_{3_0} \quad , \tag{17}$$

where $\alpha_{i_0} = g_{i_0}^2/4\pi$. Then <u>defining</u> the renormalized effective couplings at mass scale M_S by minimal subtraction in the dimensional regularization scheme (with M_S also the 't Hooft unit of mass introduced to keep the bare couplings dimensionless[24])

$$\alpha_i(M_S) \equiv \alpha_{i_0}\left[1 - \frac{\alpha_{i_0}}{\pi}\frac{C_1}{n-4} - \frac{\alpha_{i_0}^2}{\pi^2}\left(\frac{C_2}{(n-4)^2} + \frac{C_3}{n-4}\right) - O(\alpha_{i_0}^3)\right] , \qquad (18)$$

where the C_j are constants independent of i which depend on all particles, light and super-heavy.[25] Because the right hand side of (18) doesn't involve any quantities that distinguish the $\alpha_i(M_S)$, i=1,2,3 from one another, they are exactly equal

$$\alpha_1(M_S) = \alpha_2(M_S) = \alpha_3(M_S) \quad . \qquad (19)$$

Furthermore, the effective electromagnetic coupling at mass scale M_S is defined by

$$\frac{1}{\alpha(M_S)} \quad \frac{5}{3\alpha_1(M_S)} + \frac{1}{\alpha_2(M_S)} = \frac{8}{3\alpha_2(M_S)} \quad ; \qquad (20)$$

hence

$$\sin^2\theta_W(M_S) = \frac{\alpha(M_S)}{\alpha_2(M_S)} = \frac{3}{8} \quad . \qquad (21)$$

This procedure answers the question people often raise, where do the effective couplings $\alpha_i(\mu)$ become equal, at $\mu = M_S$, $10M_S$, $100M_S$ etc.? The answer is that because they are renormalized quantities, the unification point is largely a matter of choice. I have <u>defined</u> the effective couplings to be exactly equal at M_S. In the SU(5) model this turns out to be the most natural choice.[26] Next, using (19) as a boundary condition, one obtains effective couplings $\alpha_i(\mu)$ at mass scale μ exactly by integrating the <u>effective</u> beta functions

$$\mu \frac{\partial}{\partial\mu} \alpha_i = \beta(\alpha_i) = b_i\alpha_i^2 + b_{ij}\alpha_i^2\alpha_j \quad , \qquad i,j=1,2,3 \quad , \tag{22}$$

where b_i and b_{ij} receive contributions only from particles with masses $<< M_S$ i.e. the ordinary low mass particles of the $SU(3)_c \times SU(2) \times U(1)$ theory. (Only the first two terms in the β functions are required.) In that way one obtains

$$\frac{1}{\alpha_i(\mu)} = \frac{1}{\alpha_i(M_S)} + b_i \,\ell n \frac{M_S}{\mu} + F_i(\alpha_j(M_S), \alpha_j(\mu)), \; i,j=1,2,3 \tag{23}$$

where

$$b_1 = -\frac{1}{2\pi}\left(-\frac{4}{3}n_g - \frac{1}{10}N_H\right) \quad ,$$

$$b_2 = -\frac{1}{2\pi}\left(\frac{22}{3} - \frac{4}{3}n_g - \frac{1}{6}N_H\right) \quad , \tag{24}$$

$$b_3 = -\frac{1}{2\pi}\left(11 - \frac{4}{3}n_g\right) \quad ,$$

n_g=number of fermion generations (I take n_g=3) and N_H=number of SU(2) Higgs isodoublets with mass $\simeq M_W$. The F_i are complicated $O(\alpha)$ contributions resulting from the b_{ij} terms in (22) which I won't write out; but I do include their effect in my final results.[3] As before, the effective electromagnetic coupling is given by

$$\frac{1}{\alpha(\mu)} = \frac{5}{3\alpha_1(\mu)} + \frac{1}{\alpha_2(\mu)} \quad . \tag{25}$$

Using this formalism along with n_g=3 in (24), I find from (21), (23) and (25) (for $\mu=M_W$ the $W^{\pm}$ boson mass; a natural comparison point)

$$\frac{\alpha(M_W)}{\alpha_s(M_W)} = \frac{3}{8}\left[1 - \left(11 + \frac{N_H}{6}\right)\frac{\alpha(M_W)}{\pi}\,\ell n\,\frac{M_S}{M_W} + O\left(\frac{\alpha^2(M_W)}{\pi^2}\,\ell n\,\frac{M_S}{M_W}\right)\right] \quad , \tag{26}$$

$$\sin^2\theta_W(M_W) = \frac{3}{8}\left[1 - \left(\frac{110-N_H}{18}\right)\frac{\alpha(M_W)}{\pi}\,\ell n\,\frac{M_S}{M_W} + O\left(\frac{\alpha^2(M_W)}{\pi^2}\,\ell n\,\frac{M_S}{M_W}\right)\right] \quad , \tag{27}$$

where[8,13]

$$M_W \simeq \frac{38.5}{\sin\theta_W(M_W)}\ \text{GeV} \quad . \tag{28}$$

(I subsequently take N_H=1 for illustrative purposes.) All higher order terms in (26) and (27) are exactly obtainable from the F_i in (23), i.e. they originate from the b_{ij} terms in (22). The nice features of this procedure are: 1) (26) and (27) are exact when the b_{ij} contributions are retained 2) There are no threshold effects. 3) There are no $O(\alpha)$ nonlogarithmic corrections to the relationships in (26) and (27). One might assume that these nice features are offset by resulting complications in obtaining the quantities $\alpha(M_W)$, $\alpha_s(M_W)$ and $\sin^2\theta_W(M_W)$ from present day experiments; however, that is not the case. Because minimal subtraction is employed in the definition of the renormalized couplings in (18), the $\alpha_i(M_W)$ turn out to be simply related to quantities being presently measured experimentally. For example, $\sin^2\theta_W(M_W)$ in (27) is essentially the quantity being presently measured in neutral current experiments (see eq. (15)), modulo very small radiative corrections of $O(\alpha)$ which must be applied to those experiments in any case.[13,27] The effective electromagnetic coupling $\alpha(M_W)$ is related to the usual fine structure constant (α=1/137) by[8,13,22]

$$\alpha(M_W) \simeq \alpha\left[1 + \frac{2\alpha}{3\pi}\sum_f Q_f^2\,\ell n\,\frac{M_W}{m_f}\right] \simeq \frac{1}{128.5} \quad , \tag{29}$$

where the sum is over all fermions (a factor of three for colored quarks), m_f their mass and Q_f their electric charge. (Care must be exercised in the hadronic sector, see ref. 8.) The estimate in (29) has been previously described;[13] the uncertainty in $\alpha^{-1}(M_W)$ is about ±0.5. The effective strong coupling $\alpha_s(M_W)$ is obtained from the usually quoted quantity $\alpha_s(\sqrt{10}$ GeV) using two terms in $\beta(\alpha_s)$. I find for $m_b \simeq 4.5$ GeV and $m_t \simeq 18$ GeV

$$\alpha_s(\sqrt{10}\ \text{GeV})=0.30 \rightarrow \alpha_s(M_W) \simeq 0.134 \qquad \left(\begin{array}{c}\text{Electroproduction} \\ \Lambda_{\overline{MS}} = 0.5\ \text{GeV}\end{array} \right), \tag{30}$$

$$\alpha_s(\sqrt{10}\ \text{GeV})=0.20 \rightarrow \alpha_s(M_W) \simeq 0.110 \qquad (\text{Charmonium}) \quad . \tag{31}$$

The value of α_s in (30) comes from the analysis of scaling violations in electroproduction while (31) follows from charmonium spectroscopy. Notice that a rather large uncertainty in α_s at $\mu = \sqrt{10}$ GeV becomes a much smaller uncertainty at $\mu = M_W$, a fortunate occurrence for this analysis.

Solving (26) or (27) for M_S, one obtains the following leading order expressions

$$M_S^{(1)} \simeq M_W \exp\left[\frac{6\pi}{66+N_H} \frac{1}{\alpha(M_W)} \left(1 - \frac{8\alpha(M_W)}{3\alpha_s(M_W)}\right)\right] \quad , \tag{32}$$

$$M_S^{(1)} \simeq M_W \exp\left[\frac{18\pi}{110-N_H} \frac{1}{\alpha(M_W)} \left(1 - \frac{8}{3}\sin^2\theta_W(M_W)\right)\right] \quad , \tag{33}$$

where the superscript (1) indicates that only the first term b_i in each beta function has been employed (that turns out to be a good approximation, especially in (33)). One also finds (to leading order)

$$\sin^2\theta_W(M_W) \quad \frac{11+\frac{1}{2}N_H}{66+N_H}\left(1 + \frac{\alpha(M_W)}{\alpha_s(M_W)}\ \frac{110-N_H}{33+\frac{3}{2}N_H}\right) \tag{34}$$

(the b_{ij} terms in (22) increase this prediction for $\sin^2\theta_W(M_W)$ by about 2%). It is noteworthy that increasing N_H affects (32) and (33) in opposite ways. This feature can be used to bound N_H.

If the second terms (the b_{ij}) in the β functions of (22) are included, they modify the super-heavy mass predictions of (32) and (33) in approximately the following way

$$M_S^{(2)} \simeq M_S^{(1)} \times (.92) \left(\frac{\alpha_s(M_S)}{\alpha_s(M_W)}\right)^{144/(462+7N_H)} \simeq \frac{1}{2} M_S^{(1)} \tag{35}$$

for the M_S value coming from (32) and

$$M_S^{(2)} \simeq M_S^{(1)} \times (1.03) \left(\frac{\alpha_s(M_S)}{\alpha_s(M_W)}\right)^{-24/(7770-7N_H)} \simeq 1.09 M_S^{(1)} \tag{36}$$

for (33). [This analysis is for $n_g=3 \rightarrow \alpha_i(M_S) \simeq 0.024$.] Note that the higher order effects are not as significant for the predicted value of M_S obtained using $\sin^2\theta_W(M_W)$ as input, i.e. eq. (36), as they are when $\alpha_s(M_W)$ is employed as in eq. (35).

To illustrate the SU(5) model's predictions, I have listed in Table II values of $\sin^2\theta_W(M_W)$, M_S and τ_p [28] corresponding to $\alpha_s(M_W)$ in the range 0.10 - 0.20.

Table II. Values for $\sin^2\theta_W(M_W)$, M_S and τ_p predicted by the SU(5) model for a range of $\alpha_s(M_W)$.

$\alpha_s(M_W)$	$\sin^2\theta_W(M_W)$	M_S (GeV)	τ_p(yr)
0.10	0.218	1.4×10^{14}	$(0.8-15)\times10^{28}$
0.11	0.214	2.6×10^{14}	$(1.0-19)\times10^{29}$
0.12	0.211	4.5×10^{14}	$(8.2-156)\times10^{29}$
0.13	0.208	7.2×10^{14}	$(5.4-102)\times10^{30}$
0.14	0.206	1.1×10^{15}	$(2.6-50)\times10^{31}$
0.15	0.204	1.5×10^{15}	$(1.0-19)\times10^{32}$
0.16	0.202	2.0×10^{15}	$(3.2-61)\times10^{32}$
0.17	0.200	2.6×10^{15}	$(9.1-174)\times10^{32}$
0.18	0.199	3.3×10^{15}	$(2.4-45)\times10^{33}$
0.19	0.198	4.1×10^{15}	$(5.7-107)\times10^{33}$
0.20	0.197	4.9×10^{15}	$(1.2-22)\times10^{34}$

The values given in Table II were obtained using $N_H=1$, $\alpha(M_W) = 1/128.5$ and the formulas I have presented including the b_{ij} in the β functions. For $\alpha_S(M_W)=0.134$ as suggested by electroproduction experiments, $\sin^2\theta_W(M_W)=0.207$, $M_S \simeq 8.5\ 10^{14}$ GeV and $\tau_p \simeq 1-20\times10^{31}$ yrs; these values I take as at present the best SU(5) model predictions. For $\Lambda_{\overline{MS}} = 0.4$ GeV $\rightarrow \alpha_s(M_W) \simeq 0.127$, the value used by Goldman and Ross,[22] I find $M_S \simeq 6.3 \times 10^{14}$ GeV, which is in rather good agreement with the result in (14). If $\alpha_s(M_S)=0.11$ as suggested by charmonium spectroscopy, then $\sin^2\theta_W(M_W)$ is predicted to be 0.214 (better agreement with experiment) and τ_p should be right at the present experimental bound. (Perhaps the truth lies between these two sets of predictions.)

Can N_H be very large? The answer is no. For example if N_H = 10 and $\alpha_s(M_W)$ = 0.14, I find

$$\sin^2\theta_W(M_W) \simeq 0.24 \;, \quad M_S \simeq 2.8 \times 10^{13} \text{ GeV}, \quad \tau_p \simeq (1.3\text{-}25)\times 10^{25} \text{ yr} \tag{37}$$

So increasing N_H pushes $\sin^2\theta_W(M_W)$ closer to experiment; but M_S and τ_p quickly become unacceptably small. Perhaps $N_H \simeq 3$ would give optimal predictions.

IV. SUMMARY

I have exhibited the results of a new procedure for extracting M_S from radiative corrections in the SU(5) model. This approach avoids complications due to threshold effects. For $\alpha(M_W)$=1/128.5, N_H=1 and $\alpha_s(M_W)$=0.134 as suggested by electroproduction experiments, I find

$$\sin^2\theta_W(M_W) \simeq 0.207 \;, \quad M_S \simeq 8.5 \times 10^{14} \text{ GeV} \;, \quad \tau_p \simeq 1\text{-}20\times 10^{31} \text{ yr.} \tag{38}$$

The proton lifetime prediction in (38) is somewhat longer than the estimates of earlier analyses[22] (see ref. 22); fortunately it still falls within the range of experimentally accessible values (see eq. (2)). If $\alpha_s(M_W)$=0.11 as suggested by charmonium spectroscopy, then $\tau_p \simeq 1\text{-}20\times 10^{29}$ yr right at the present bound (see eq. (1)).

There is still considerable uncertainty in the theoretical estimate of τ_p. Some remaining important questions are: Which calculation of τ_p represents the better estimate, (12) or (13)? Remember, they differ by a factor of 19. Is N_H>1? If so, the predicted value of τ_p decreases while $\sin^2\theta_W(M_W)$ increases; an interesting possibility. What is the appropriate value of $\alpha_s(M_W)$? Fortunately, as illustrated in Table II, the SU(5) model's predictions are not wildly sensitive to reasonable variations in this quantity. What will the experimental value of $\sin^2\theta_W(M_W)$ become after radiative corrections are accounted for? Those calculations will be forthcoming shorthly.[27] Perhaps most importantly, is $\alpha(M_W)=1/128.5$ a good estimate? If in fact I had used $\alpha(M_W)=1/129$, then for $\alpha_s(M_W)=0.14$ and $N_H=1$. I find

$$\sin^2\theta_W(M_W)\simeq 0.206, \quad M_S\simeq 1.23\times 10^{15}\ \text{GeV}, \quad \tau_p\simeq(4.6\text{-}88)\times 10^{31}\text{yr} \tag{39}$$

So unless my estimated uncertainty in $\alpha(M_W)$ is way off, the quoted results don't seem to depend heavily on using $\alpha(M_W)=1/128.5$. (Although using $\alpha(M_W)\simeq 1/137$ would have erroneously overestimated M_S by an order of magnitude.)

All of the above theoretical questions are interesting. However, the most important question: Can the proton actually decay? will only be answered by experiment.

REFERENCES

1. M. Goldhaber, proceedings of this conference.
2. H. Georgi and S. Glashow, Phys. Rev. Lett. 32, 438 (1974).
3. W. Marciano, to be published.
4. F. Reines and M. Crouch, Phys. Rev. Lett. 32, 493 (1974); J. Learned, F. Reines and A. Soni, Phys. Rev. Lett. 43, 907 (1979).
5. For a review of proposed proton decay searches, see L. Sulak, Proceedings of the International Conference on Neutrino Physics, Bergen 1979.
6. For a review of grand unified theories and the question of proton stability, see M. Gell-Mann, P. Ramond and R. Slansky, Rev. Mod. Physics 50, 721 (1978).
7. A. Buras, J. Ellis, M. Gaillard, and D. Nanopoulos, Nucl. Phys. B135, 66 (1978).
8. W. Marciano, Phys. Rev. D20, 274 (1979).
9. Properties of the SU(5) model have been reviewed by D. Nanopoulos, Protons Are Not Forever, Harvard preprint (1978);

J. Ellis, SU(5), CERN preprint (1979).

10. M. Chanowitz, J. Ellis and M. Gaillard, Nucl. Phys. B128, 506 (1977).

11. S. Weinberg, Phys. Rev. Lett. 19, 1264 (1967); A. Salam in Elementary Particle Physics, edited by N. Svartholm (Almqvist and Wiksells, Stockholm, 1968), p. 367.

12. H. Georgi, H. Quinn and S. Weinberg, Phys. Rev. Lett. 33, 451 (1974). The existence of large renormalization effects in the SU(5) model and a procedure for calculating them was demonstrated in this pioneering work. The results I present are refinements of the method described by these authors.

13. W. Marciano, Invited talk at the APS Division of Particles and Fields Meeting, McGill University, October 1979; Rockefeller University preprint COO-2232B-192. This talk explains in some detail the effect of radiative corrections on experimental determinations of $\sin^2\theta_W$ and continues the analysis of ref. 8. Note α_{10} in that talk is 3/5 times the α_{10} used here.

14. M. Yoshimura, Phys. Rev. Lett. 41, 281 (1978); 42, 746 (E) 1979.

15. M. Machacek, Nucl. Phys. B159, 37 (1979). This paper provides a detailed description of proton decay in the SU(5) model and some of its generalizations.

16. S. Weinberg, Phys. Rev. Lett. 43, 1566 (1979); F. Wilczek and A. Zee, Phys. Rev. Lett. 43, 1571 (1979).

17. Detailed descriptions of the ingredients that go into calculating τ_p are given in refs. 7 and 15.

18. J. Ellis, M. Gaillard and D. Nanopoulos, CERN preprint TH. 2749 (1979). This paper describes the SU(2)×U(1) enhancement factors.

19. C. Jarlskog and F. Yndurain, Nucl. Phys. B149, 29 (1979).

20. A. Din, G. Girardi and P. Sorba, Lapp preprint TH-08 (1979).

21. J. Donoghue, MIT preprint CTP #824 (1979).

22. T. Goldman and D. Ross, Phys. Lett. 84B, 208 (1979); preprint CALT-68-759 (1980). The latter of these gives the estimate in eq. (14).

23. P. Langacker, J. Kim, M. Levine, H. Williams and D. Sidhu, Univ. of Penn. preprint COO-3071-243 (1979).

24. W. Marciano, Phys. Rev. D12, 3861 (1975).

25. In addition to poles, I also subtract extraneous contributions such as Euler's constant γ and $\sqrt{4\pi}$ which arise in dimensional regularization. That is why I use the QCD parameter $\Lambda_{\overline{MS}}$ extracted from experiment.

26. S. Weinberg, Effective Gauge Theories, Harvard preprint HUTP-80/A001 (1980). In this paper a general formalism for integrating out the contribution of heavy particles is given. It provides a simple prescription for extending the SU(5) analysis to theories with more than one super-heavy mass scale.

27. W. Marciano and A. Sirlin, to be published.

28. The predictions for τ_p in eqs. (12) and (13) along with the values quoted in Table I were obtained using amplitude enhancement factors appropriate for $\alpha_s(M_W)=0.14$. For values of $\alpha_s(M_W)$ smaller than 0.14, the proton lifetime τ_p is actually smaller than the results I have given, because the enhancement factors should be increased; the opposite effect occurs for $\alpha_s(M_W)>0.14$.

FUN WITH E_6

R. Slansky

Los Alamos Scientific Laboratory*

University of California, Los Alamos, New Mexico 87545

ABSTRACT

The exceptional Lie group E_6 is a candidate local symmetry for a Yang-Mills theory that unifies electromagnetic, weak, and strong interactions. Several ways of incorporating the fermion spectrum are discussed, including an amusing example where some of the known spin 1/2 fermions are composite states of elementary fermions and some scalar particles in the theory. The symmetry properties and the representations of E_6 are reviewed, the symmetry breaking classified, and the dynamical breaking of the weak interaction gauge group is discussed, all in some detail using Dynkin's representation theory.

I. INTRODUCTION

It does not look likely that it should be much fun to explore the features of a Yang-Mills theory based on a group of rank 6 with 78 generators. Just writing down the commutation relations of the exceptional Lie algebra E_6 in a useful basis looks like a mess, and working with representations of dimension 27, 78, 351, or higher would seem to be, at best, tedious. The only way to have any

---

*Work supported by the U.S. Department of Energy.

fun at all in exploring the consequences of such a big group is to have an easily applied technology. In fact, Dynkin's representation theory[1] is quite adequate for obtaining many of the symmetry results of the theory. The purpose of this talk is twofold: (1) to "review" how these techniques can be used to explore certain consequences of Yang-Mills theories; and (2) to apply these techniques to an E_6 model in which the scalars in an adjoint $\underline{78}$ form composite bound states with the elementary fermions in a $\underline{27}$ to give a total of three quark-lepton families. Much of the research reported here was done in collaboration with Gordon Shaw.[2] I have written up these notes in an order reversed from the talk at Orbis Scientiae and the presentation of Ref. 2, because I would like to use this opportunity to emphasize the usefulness of Dynkin's representation theory for exploring unified models. Instead of starting with the physics of the composite fermion model, which the reader will find in Sec. 3, I will review some of the mathematics of E_6 first, beginning with a rapid review of Dynkin's representation theory. We can then use the language developed in Sec. 2 with impunity for describing the composite fermion model.

II. SYMMETRY PROPERTIES OF E_6

Our object in this Sec. is to review the quantum number structure of the representation of E_6 in a fashion that is very convenient for many other applications. This means that we must be able to identify the physical significance and compute the eigenvalues of the 6 diagonalizable generators of E_6, when acting on vectors in the representation, and then learn what the 72 ladder operators do to these quantum numbers. The Dynkin formalism works for any simple algebra. It can be viewed as a fancy way for keeping track of the quantum numbers, and as such, it is merely a mathematical bookkeeping trick. However, it does greatly simplify doing the physics too, and I hope the reader will find it to be fun.

The rank of an algebra G is the number of independent diagonalizable generators in G; these ℓ generators form the Cartan subalgebra of G. The remaining generators can be written as ladder operators, which, when acting on a Hilbert space vector in a representation, change the set of ℓ eigenvalues by amount α. The ℓ eigenvalues correspond to a point in an ℓ dimensional Euclidean space, which is called the weight of the representation vector Λ. The root α is also a vector with ℓ components in weight space, and it represents a permissible shift from one weight to another. If Λ and $\Lambda + \alpha$ are both weights of a unitary irreducible representation (irrep) of G, the the ladder operator E_α acting on $|\Lambda>$ is proportional to $|\Lambda + \alpha>$. The correspondance between weight space and representation space is a key point in representation theory.[1,3]

A convenient basis for the weight space is formed by the ℓ simple roots of the algebra; these specially chosen roots have the relative lengths and angles indicated by the Dynkin diagram.[1] The simple root basis is not an orthonormal basis for weight space, but otherwise, it is extremely useful. Each representation vector is labeled by a weight. In the "Dynkin basis" the components of a weight Λ are ℓ integers a_i, defined by the scalar products,

$$a_i = \frac{2(\Lambda,\alpha_i)}{(\alpha_i,\alpha_i)} \quad , \quad i = 1,2,\ldots,\ell, \tag{1}$$

where α_i is the i-th simple root. The proof that a_i must be an integer is a generalization of the proof that the magnetic quantum number of an SU_2 vector must be integer or half integer. (For SU_2, $a_1 = 2m$.) The computation of the set of weights of an irrep written in the Dynkin basis $(a_1a_2\ldots a_\ell)$ is straightforward. There exists one weight, the highest weight, that always labels a unique Hilbert space vector in the irrep. Each set of ℓ nonnegative integers uniquely gives the highest weight of an irrep of G, and these sets exhaust the entire set of finite dimensional irreps. Starting from the highest weight, the remaining weights are computed by subtracting off simple roots: if at any level (the level is the number of simple roots that have been subtracted from the highest weight) the a_i coefficient is positive, then the i-th simple root can be subtracted off a_i more times, and the resulting weight in the irrep. The maximum number of simple roots that can be subtracted off the highest weight is the scalar product of the highest weight and level vector; the level vector for E_6 is $\tilde{R} = [16,30,42,30,16,22]$. (The computation of scalar products will be considered shortly.) Moreover, the weight system of the representation must be spindle-shaped; the number of weights at level k is equal to the number of weights at level $\tilde{R}\cdot\Lambda - k$ (Λ the highest weight), and the number of weights at level k+1 is $\geq$ to that at level k, for k less than $\tilde{R}\cdot\Lambda/2$. A given weight may be obtained by several routes. The degeneracy of a weight can be computed (with a bit more difficulty) from the Freudenthal recursion relation, but for the low lying irreps studied here, the degeneracy is easily guessed. (The degeneracy is the number of Hilbert space vectors in an irrep with the same weight; additional labels are needed to distinguish degenerate weights.)

We will need to compute many scalar products in weight space. Because the simple root basis is not orthonormal, the scalar product of weights with components a_i, a'_i involves a metric tensor, which is closely related to the inverse of the Cartan matrix A^{-1}, and is A^{-1} for the algebras where the simple roots all have the same lengths. Thus, the scalar product of Λ and Λ' is

$$(\Lambda, \Lambda') = \sum_{ij=1}^{\ell} a_i (A^{-1})_{ij} a'_j \equiv \tilde{\Lambda}\cdot\Lambda = \sum_{i=1}^{\ell} \tilde{a}_i a'_i , \qquad (2)$$

$$\tilde{a}_i = a_j (A^{-1})_{ji} .$$

We will often give Λ in the "Dynkin basis" $(a_1 \ldots a_\ell)$, and we will call it $\tilde{\Lambda}$ when multiplied by the metric, $\tilde{\Lambda} = [\tilde{a}_1 \ldots \tilde{a}_\ell]$, putting the components in square brackets. The inverse of the Cartan matrix (A itself can be read off the Dynkin diagram) for E_6 is

$$A^{-1} = \frac{1}{3}\begin{pmatrix} 4 & 5 & 6 & 4 & 2 & 3 \\ 5 & 10 & 12 & 8 & 4 & 6 \\ 6 & 12 & 18 & 12 & 6 & 9 \\ 4 & 8 & 12 & 10 & 5 & 6 \\ 2 & 4 & 6 & 5 & 4 & 3 \\ 3 & 6 & 9 & 6 & 3 & 6 \end{pmatrix} . \quad (3)$$

Let us apply these results to some E_6 irreps. The representation with highest weight (1 0 0 0 0 0) is complex and $\underline{27}$ dimensional. With Dynkin's conventions, the first simple root is (2-1 0 0 0 0), so the first level weight is the (-1 1 0 0 0 0); the second level is (0-1 1 0 0 0); the third level is (0 0-1 1 0 1); the fourth level has two weights, (0 0 0-1 1 1) and (0 0 0 1 0-1); and so on to the 16-th level, (0 0 0 0-1 0). The $\overline{\underline{27}}$ has highest weight (0 0 0 0 1 0), and the adjoint $\underline{78}$ has highest weight (0 0 0 0 0 1).

We shall show shortly that in the usual embedding of color and flavor in E_6, the electric charge operator, which is in the Cartan subalgebra, is measured along the axis,

$$\tilde{Q}^{em} = \frac{1}{3}[\ 2\ 1\ 2\ 0\ 1\ 0\ ], \quad (4)$$

where the normalization of $\tilde{Q}^{em}$ is chosen so that the electric charge of any weight (or state) is given by the scalar product,

$$Q^{em}_{\Lambda} = \tilde{Q}^{em} \cdot \Lambda \quad . \quad (5)$$

Thus, the (1 0 0 0 0 0) has electric charge 2/3, (-1 1 0 0 0 0) has charge -1/3, (0 0-1 1 0 1) has charge -2/3, and so on. We will need to discuss the conventions concerning the embedding of electromagnetism in E_6.

Before we can identify the physical relevance of the roots

and axes in weight space, we must find out how color and flavor are embedded in E_6. This embedding can be done in a coordinate independent fashion.[4,5] There is only one embedding of QCD and QED that seems to have a chance of being relevant. This embedding is identified by the requirement that the 27 has 9 color singlets, 3 quarks and 3 antiquarks, where two of the quarks have electric charge -1/3 and the other quark has charge 2/3. (The embeddings with one singlet, one octet, 3 quarks, and 3 antiquarks appear irrelevant, as do more exotic charge assignments.) For many purposes this coordinate independent statement of the embedding is sufficient. However, for practical calculations of symmetry breaking and mass matrices, it is often helpful to have a coordinatization of the weight space. We follow a certain set of conventions here; other conventions are related by a Weyl reflection.

The possible subgroup chains that lead to these color and charge assignments have been classified by Dynkin.[1] The most useful for physics are

$$E_6 \supset SO_{10} \supset SU_5 \supset SU_2^{W} \times SU_3^{C} \ , \qquad (6)$$

$$E_6 \supset SO_{10} \supset SU_2^{W} \times SU_2 \times SU_4 \ (SU_4 \supset SU_3^{C}) \ , \qquad (7)$$

$$E_6 \supset SU_3 \times SU_3 \times SU_3^{C} \ , \qquad (8)$$

where the first SU_3 in (8) contains SU_2^{W}, the weak isospin. (We have ignored the U_1's, but this omission will be filled in later.) Our embedding conventions are to follow Ref. 6, where the highest weights of an irrep are projected onto highest weights of the irreps to which it branches, for the chain in (6). Then for (7) and (8), we require that the same physical directions in E_6 weight space as derived from (6) are maintained for the other embeddings. (For example, the same roots correspond to SU_3^{C}, etc.) The projection matrices for the subgroup chain in (6) are[6]

$$P(\ E_6 \supset SO_{10}) = \begin{pmatrix} 0 & 1 & 1 & 1 & 0 & 0 \\ 0 & 0 & 0 & 0 & 0 & 1 \\ 0 & 0 & 1 & 0 & 0 & 0 \\ 0 & 0 & 0 & 1 & 1 & 0 \\ 1 & 1 & 0 & 0 & 0 & 0 \end{pmatrix}, \qquad (9)$$

$$P(\ SO_{10} \supset SU_5) = \begin{pmatrix} 1 & 1 & 0 & 0 & 0 \\ 0 & 0 & 1 & 0 & 1 \\ 0 & 0 & 0 & 1 & 0 \\ 0 & 1 & 1 & 0 & 0 \end{pmatrix}, \qquad (10)$$

$$P(\ SU_5 \supset SU_2 \times SU_3) = \begin{pmatrix} 0 & 1 & 1 & 0 \\ 1 & 1 & 0 & 0 \\ 0 & 0 & 1 & 1 \end{pmatrix}. \qquad (11)$$

The projection matrix is not square if there is a loss of rank or a U_1 factor in going to the subgroup. The matrix elements are nonnegative integers because of the conventions followed.

We identify the SU_2 and SU_3 of (11) as the weak and color groups, and use this subgroup chain to identify the physical significance of the roots of E_6. This identification is worked out in Table 1. Consider the example of the E_6 root (1-1 1-1 1 0), which is projected onto (-1 0 1 0 0) by (9). That is a root in the $\underline{45}$ of SO_{10}. This SO_{10} weight is then projected by (10) to the SU_5 weight (-1 1 0 1), which is a root in the adjoint $\underline{24}$. Finally, (11) projects (-1 1 0 1) to (1) (0 1) of $SU_2 \times SU_3$, which identifies the (1-1 1-1 1 0) root of E_6 as a color triplet with $I_3^w = 1/2$; it is the charge 4/3, SU_5 antilepto-diquark that mediates proton decay. It is a simple computation to construct the rest of Table 1 for the $\underline{78}$, and also to work out Table 2 for the $\underline{27}$ of E_6. The columns labeled $SU_5(SO_{10})$ gives the SO_{10} irrep into which the E_6

Table 1. Nonzero E_6 Roots.

root	level	color	Q^{em}	I_3^w	Q^t	$SU_5(SO_{10})$	$\tilde{B}.\alpha$	$\tilde{L}.\alpha$
Color SU_3 roots								
(0 0 0 0 0 1)	0	(1 1)	0	0	0	24(45)	0	0
(0 1 0 0-1 0)	4	(2-1)	0	0	0	24(45)	0	0
(0-1 0 0 1 1)	7	(-1 2)	0	0	0	24(45)	0	0
Left-handed SU_3 roots								
(1 0 0 0 1-1)	6	(0 0)	1	1	0	24(45)	0	± 1
(-1 1 0 0 1-1)	7	(0 0)	0	1/2	-3	$\bar{5}(16)$	3c	$3d\pm 2$
(-2 1 0 0 0 0)	12	(0 0)	-1	-1/2	-3	$\bar{5}(16)$	3c	$3d\pm 1$
Right-handed SU_3 roots								
(0-1 1 1-1-1)	9	(0 0)	0	0	3	$1(\overline{16})$	a+b-2c	$-d+2e\mp 2$
(0 0-1 2-1 0)	10	(0 0)	-1	0	3	$\overline{10}(\overline{16})$	-a+2b-c	$-2d+e\mp 1$
(0-1 2-1 0-1)	10	(0 0)	1	0	0	10(45)	2a-b-c	$d+e\mp 1$
SU_5 antilepto-diquarks								
(1-1 1-1 1 0)	4	(0 1)	4/3	1/2	0	24(45)	a-b-c	0
(1 0 1-1 0-1)	8	(1-1)	4/3	1/2	0	24(45)	a-b-c	0
(1-1 1-1 1-1)	15	(-1 0)	4/3	1/2	0	24(45)	a-b-c	0
(0-1 1-1 0 1)	9	(0 1)	1/3	-1/2	0	24(45)	a-b-c	∓ 1
(0 0 1-1-1 0)	13	(1-1)	1/3	-1/2	0	24(45)	a-b-c	∓ 1
(0-1 1-1 0 0)	20	(-1 0)	1/3	-1/2	0	24(45)	a-b-c	∓ 1

SO_{10}/SU_5 leptoquarks								
(0 0 1 0 0-1)	1	(1 0)	2/3	1/2	0	10(45)	a	d+e
(0-1 1 0 1-1)	8	(-1 1)	2/3	1/2	0	10(45)	a	d+e
(0 0 1 0 0-2)	12	(0-1)	2/3	1/2	0	10(45)	a	d+e
(-1 0 1 0-1 0)	6	(1 0)	-1/3	-1/2	0	10(45)	a	d+e∓1
(-1-1 1 0 0 0)	13	(-1 1)	-1/3	-1/2	0	10(45)	a	d+e∓1
(-1 0 1 0-1-1)	17	(0-1)	-1/3	-1/2	0	10(45	a	d+e∓1
(-1 0 0 1 0 0)	4	(0 1)	-2/3	0	0	10(45)	b+c	d+e
(-1 1 0 1-1-1)	8	(1-1)	-2/3	0	0	10(45)	b+c	d+e
(-1 0 0 1 0-1)	15	(-1 0)	-2/3	0	0	10(45)	b+c	d+e
E_6/SO_{10} leptoquarks								
(0 1 0-1 1 0)	3	(1 0)	2/3	1/2	-3	10(16)	-b+2c	2d-e±2
(0 0 0-1 2 0)	10	(-1 1)	2/3	1/2	-3	10(16)	-b+2c	2d-e±2
(0 1 0-1 1-1)	14	(0-1)	2/3	1/2	-3	10(16)	-b+2c	2d-e±2
(-1 1 0-1 0 1)	8	(1 0)	-1/3	-1/2	-3	10(16)	-b+2c	2d-e±1
(-1 0 0-1 1 1)	15	(-1 1)	-1/3	-1/2	-3	10(16)	-b+2c	2d-e±1
(-1 1 0-1 0 0)	19	(0-1)	-1/3	-1/2	-3	10(16)	-b+2c	2d-e±1
(-1 0 1-1 1 0)	5	(0 1)	1/3	0	-3	$\bar{5}$(16)	a-b+2c	3d±1
(-1 1 1-1 0-1)	9	(1-1)	1/3	0	-3	$\bar{5}$(16)	a-b+2c	3d±1
(-1 0 1-1 1-1)	16	(-1 0)	1/3	0	-3	$\bar{5}$(16)	a-b+2c	3d±1
(-1 1-1 0 1 1)	6	(0 1)	-2/3	0	-3	10(16)	-a+3c	2d-e±2
(-1 2-1 0 0 0)	10	(1-1)	-2/3	0	-3	10(16)	-a+3c	2d-e±2
(-1 1-1 0 1 0)	17	(-1 0)	-2/3	0	-3	10(16)	-a+3c	2d-e±2

weight branches, and then tells which SU_5 irrep that the SO_{10} weight branches into.[7] The simple roots are recovered at level 10 or minus level 12. In computing the projection matrices for the subgroup chains (7) and (8), we require that the E_6 roots have the same interpretation as the chain given by (9), (10), and (11). That is our convention. Thus we must require that the projection matrix for (8) carries the (1-1 1-1 1 0) root to (1 0)(1 0)(0 1) of $SU_3 \times SU_3 \times SU_3{}^c$, since this root has $I_3{}^w = 1/2$, $Q^{em} = 4/3$, and is an antiquark. Since the $\underline{78}$ branches to $(\underline{8},\underline{1},\underline{1}) + (\underline{1},\underline{8},\underline{1}) + (\underline{1},\underline{1},\underline{8}) + (\underline{3},\underline{3},\overline{\underline{3}}) + (\overline{\underline{3}},\overline{\underline{3}},\underline{3})$, we see that the identification is unique. For the subgroup chain in (7), we use (9) and find the projection matrices,

$$P(\ SO_{10} \supset SU_2 \quad SU_2 \times SU_4) = \begin{pmatrix} 0 & 0 & 1 & 1 & 1 \\ 0 & 0 & 1 & 0 & 0 \\ 1 & 1 & 1 & 0 & 1 \\ 0 & 1 & 1 & 1 & 0 \\ -1 & -1 & -1 & -1 & 0 \end{pmatrix} , \qquad (12)$$

$$P(\ SU_4 \supset SU_3{}^c) = \begin{pmatrix} 1 & 0 & 0 \\ 0 & 1 & 0 \end{pmatrix} . \qquad (13)$$

Of course, (12) and (13) are useful for studying SO_{10} theories. Finally, we find for the projection matrix of (8),

$$P(\ E_6 \supset SU_3 \times SU_3 \times SU_3^c = \begin{pmatrix} 1 & 1 & 1 & 1 & 1 & 0 \\ 0 & -1 & -1 & -1 & -1 & 0 \\ 0 & 0 & 1 & 0 & 0 & 0 \\ 0 & 0 & -1 & -1 & 0 & 0 \\ 1 & 2 & 2 & 1 & 0 & 1 \\ 0 & 0 & 1 & 1 & 1 & 1 \end{pmatrix} . \qquad (14)$$

For SU_5 theories, this machinery is stronger than is usually needed, although it is very easy to use and actually simplifies many computations. (As an exercise, consider the SU_5 fermion mass matrix including the $\underline{45}$ and $\underline{50}$ terms; remember to use the Kronecker products in doing this.)

The eigenvalues of the generators in the Cartan subalgebra can be constructed and computed in a straightforward fashion by computing the axes and normalization as done for the electric charge in (4). Since the values of these quantum numbers are already known from coordinate independent methods,[5] the computation is well defined. In Table 3 we have listed a complete set of generators for the flavor interactions. Two of the Cartan subalgebra members are in $SU_3{}^c$, and the other 4 include: The U_1 in $E_6 \supset SO_{10} \times U_1$, which we call $U_1{}^t$; the U_1 in $SO_{10} \supset SU_5 \times U_1$, which we call $U_1{}^r$; and the weak isospin and hypercharge axes. These latter axes are normalized in the usual way, so $Q^{em} = I_3{}^W + Y^W/2$.

The quantum number structure of the E_6 generators and irreps is summarized in Tables 1, 2, and 3. In Table 1, half of the nonzero roots of E_6 are listed; the negatives of those listed and the zero roots are not listed. The $\underline{27}$ is complex, so the negative of a weight in the $\underline{27}$ is not in the $\underline{27}$; it is in the $\overline{\underline{27}}$. Table 3 contains projection axes and the weights of several important roots. I hope that the reader will enjoy checking these tables and finding just how simple it is to identify the physical content of the E_6 generators and states. These techniques can be used for even bigger groups, such as SO_{18} or SO_{22}, and for even larger irreps, such as the $\underline{351}$ or $\underline{1728}$ of E_6[8] without using a computer, although a program does exist.[6]

We now turn to an important application of the above formalism, that of investigating the symmetry breaking of E_6. Instead of trying to set up a Higgs potential and doing a messy minimization problem, we apply several physical constraints that any symmetry breaking mechanism must satisfy if the standard model is to be recovered in its usual form.

If a field or "effective field" has a nonzero vacuum expectation value, then the weight of that field determines much about the symmetry breaking. E_6 models have a set of lepto-diquark bosons that mediate proton decay in second order, so these bosons must have superlarge ($\gtrsim 10^{14}$ GeV) masses from a big symmetry breaking, just as in the SU_5 model.[9] The direction of the big breaking in the 6-dimensional weight space, written in the Dynkin basis (1), is called B. Similarly, the little breaking, which gives masses to the weak bosons and may also contribute to other bosons in the theory, is called L. Of course there may be some intermediate mass scales, but generalizations to that case will be obvious.

The weight-space direction of B must be perpendicular to the $SU_3{}^c$ roots, or else the color-changing gluons will acquire masses, and it must be perpendicular to $I_+{}^w$, the root of the weak isospin raising operator, so that the charged, weak boson does not get a superlarge mass. In Table 1 we find that the $SU_3{}^c$ roots lie in the plane formed by (0 0 0 0 0 1) and (0 1 0 0-1 0), which implies that all symmetry breaking directions must have the parameterization, $\tilde{B} = [-c\ d\ a\ b\ d\ 0]$, where $\tilde{B}$ is defined in (2). The $I_+{}^w$ weight is (1 0 0 0 1-1), so the big breaking has the form,

$$\tilde{B} = [-c\ c\ a\ b\ c\ 0] \quad . \tag{15}$$

The column labeled $\tilde{B}\cdot\alpha$ in Table 1 gives a parameterization of the vector boson mass eigenvalues for each root in terms of B, up to an isoscalar factor that must be included for other than adjoint breaking.

In order to show that the boson mass eigenvalues are proportional to the weight-space scalar products of the roots and the symmetry breaking direction, we examine a simple example. The boson mass matrix in the tree approximation to the Higgs model has the form,

$$M_{\alpha\beta} = g^2 \sum_{\lambda\lambda'\lambda''} \phi_V^+(\bar{\underline{r}},-\lambda')<\bar{\underline{r}},-\lambda'|X_{-\alpha}|\bar{\underline{r}},-\lambda><\underline{r},\lambda|X_\beta|\underline{r}\lambda''>\phi_V(\underline{r},\lambda''), \quad (16)$$

where the vacuum expectation value of the scalar fields $\phi_V(r,\lambda)$ has weight λ and belongs to representation $\underset{\sim}{r}$, and $<r,\lambda|X_\alpha|\underset{\sim}{r},\lambda'>$ is a matrix element of the generator X_α. We use this notation in order to emphasize that $\phi_V(r,\lambda)$ is a tensor operator; for example, we can use the commutation relations for tensor operators to rewrite (16) as

$$M_{\alpha\beta} = -g^2 \sum_\lambda \mathrm{Tr}\{[X_{-\alpha},\phi_V^+(\bar{\underline{r}},-\lambda)][X_\beta,\phi_V(\underline{r},\lambda)]\} \quad (17)$$

Usually, (16) is more useful than (17), but there is an important case where $\phi_V(\underline{r},\lambda)$ can be expanded as a simple polynomial of the generators, so that the commutators in (17) are easily evaluated.

Suppose that $\underline{r}$ is the adjoint representation of G and that there is sufficient gauge freedom to rotate $\phi_V(\underline{r},\lambda)$ into the Cartan subalgebra of G. This means that the vacuum expected value can be expanded as

$$\phi_V(\underset{\sim}{r},\lambda) = \sum_{i=1}^{\ell} \tilde{c}_i H_i \quad . \quad (18)$$

The components $\tilde{c}_i$ define the symmetry breaking direction in weight space.

The computation of the mass matrix requires a knowledge of the commutators, $[X_\alpha, H_i]$. This commutator is zero if X_α is in the Cartan subalgebra, so we can conclude immediately that ℓ vector bosons are massless. The mass matrix is diagonal if the remaining generators E_α are written in the Cartan-Weyl basis, where the raising and lowering operators satisfy eigenvalue equations of the type,

$$[H_i , E_\alpha] = \alpha_i E_\alpha \quad , \quad (19)$$

where α_i is the i-th component of the root vector α. Upon substituting (19) into (17), we obtain

$$M^2_{\alpha\beta} = g^2 \text{ Tr } \{|d_i\tilde{C}_i) \ X_\alpha \ (\beta_j\tilde{C}_j) \ X_\beta\} = g^2\delta_{\alpha\beta}(\alpha_i\tilde{C}_i)^2 \ , \qquad (20)$$

where the last step implies a normalization convention. Thus, the
mass eigenvalue of the vector boson associated with weight α is
proportional to $(c,\alpha) = \tilde{c}.\alpha$.

We return to the more general discussion. The next question
is whether B is a direction defined by a vacuum expectation value
with zero weight, or the direction is defined by nonzero weights.
We show that B must have a contribution from zero weights. If B
were due only to a nonzero weight, then the weight must be perpendicular to the electric charge axis, since QED is not broken.
The electric charge axis (4) is perpendicular to B if $a = b + c$,
which, in turn, implies that the SU_5 lepto-diquarks receive no mass
from B. The implication for the proton decay rate is obvious.
For simplicity we can assume that the big breaking has zero weight.
Then none of the bosons associated with the Cartan subalgebra get
a superlarge mass, and we see from Table 1 that, at most, B can
break E_6 to $SU_2 \times U_1 \times U_1 \times U_1 \times SU_3^c$.

The little breaking must have a component with $|\Delta I^w| = 1/2$.
It can also have a component with $|\Delta I^w| = 0$, which has the same
form as B in (15) if it has zero weight and is constrained by
$a = b + c$ if it has nonzero weight. For now we consider the
$|\Delta I^w| = 1/2$ term only, which necessarily has nonzero weight, so
each $|\Delta I^w| = 1/2$ weight must have the form,

$$\tilde{L} = [-d, \ d\pm1, \ d + e, \ e, \ d\pm1, \ 0] \quad . \qquad (21)$$

Even without an explicit model of symmetry breaking, there
are a few more comments that may prove interesting.

(1) Much of the discussion about B is not actually restricted
to 78 breaking, except that the breaking representation must have

Table 2. Weights and Content of the 27 of E_6.

weight	level	color	Q^{em}	I_3^{w}	Q^{t}	$SU_5(SO_{10})$	SO_{10} weight
(0 0 0 1 0-1)	4	(0 0)	0	1/2	1	$\bar{5}$(16)	(1-1 0 1 0)
(-1 0 0 1-1 0)	9	(0 0)	-1	-1/2	1	$\bar{5}$(16)	(1 0 0 0-1)
(1-1 1-1 0 0)	9	(0 0)	1	0	1	10(16)	(-1 0 1-1 0)
(1 0-1 0 0 1)	10	(0 0)	0	0	1	1(16)	(-1 1-1 0 1)
(0 0 1-1 1-1)	5	(0 0)	1	1/2	-2	5(10)	(0-1 1 0 0)
(-1 0 1-1 0 0)	10	(0 0)	0	-1/2	-2	5(10)	(0 0 1-1-1)
(0 1-1 0 1 0)	6	(0 0)	0	1/2	-2	$\bar{5}$(10)	(0 0-1 1 1)
(-1 1-1 0 0 1)	11	(0 0)	-1	-1/2	-2	$\bar{5}$(10)	(0 1-1 0 0)
(1-1 0 1-1 0)	8	(0 0)	0	0	4	1(1)	(0 0 0 0 0)
(1 0 0 0 0 0)	0	(1 0)	2/3	1/2	1	10(16)	(0 0 0 0 1)
(1-1 0 0 1 0)	7	(-1 1)	2/3	1/2	1	10(16)	(-1 0 0 1 0)
(1 0 0 0 0-1)	11	(0-1)	2/3	1/2	1	10(16)	(0-1 0 0 1)
(0 0 0 0-1 1)	5	(1 0)	-1/3	-1/2	1	10(16)	(0 1 0-1 0)
(0-1 0 0 0 1)	12	(-1 1)	-1/3	-1/2	1	10(16)	(-1 1 0 0-1)
(0 0 0 0-1 0)	16	(0-1)	-1/3	-1/2	1	10(16)	(0 0 0-1 0)
(-1 1 0 0 0 0)	1	(1 0)	-1/3	0	-2	5(10)	(1 0 0 0 0)
(-1 0 0 0 1 0)	8	(-1 1)	-1/3	0	-2	5(10)	(0 0 0 1-1)
(-1 1 0 0 0-1)	12	(0-1)	-1/3	0	-2	5(10)	(1-1 0 0 0)
(0 0 0-1 1 1)	4	(0 1)	1/3	0	-2	$\bar{5}$(10)	(-1 1 0 0 0)
(0 1 0-1 0 0)	8	(1-1)	1/3	0	-2	$\bar{5}$(10)	(0 0 0-1 1)
(0 0 0-1 1 0)	15	(-1 0)	1/3	0	-2	$\bar{5}$(10)	(-1 0 0 0 0)
(0-1 1 0 0 0)	2	(0 1)	1/3	0	1	$\bar{5}$(16)	(0 0 1 0-1)
(0 0 1 0-1-1)	6	(1-1)	1/3	0	1	$\bar{5}$(16)	(1-1 1-1 0)
(0-1 1 0 0-1)	13	(-1 0)	1/3	0	1	$\bar{5}$(16)	(0-1 1 0-1)
(0 0-1 1 0 1)	3	(0 1)	-2/3	0	1	10(16)	(0 1-1 1 0)
(0 1-1 1-1 0)	7	(1-1)	-2/3	0	1	10(16)	(1 0-1 0 1)
(0 0-1 1 0 0)	14	(-1 0)	-2/3	0	1	10(16)	(0 0-1 1 0)

triality zero. The only irreps of E_6 with zero weights are $\underline{78}$, $\underline{650}$, $\underline{2430}$, $\underline{2925}$, or larger.

(2) Suppose that B is due to an adjoint representation of Higgs scalars alone. The only independent Casimir invariants of E_6 are of order 2, 5, 6, 8, 9, and 12; thus the Higgs potential can depend on the length of the $\underline{78}$ only, and in the tree approximation there are no constraints on a, b, and c in (15). The one-loop corrections to the effective potential select a = 0 and b = -c, so an entire $SO_{10} \times U_1$ is left unbroken.[10] There is no reason to believe that, when bound states and other scalars are included, these radiative corrections would dominate the determination of B. In fact, it is conceivable that B is singled out by the weak breaking.

(3) If the adjoint breaking is along the only root with $|\Delta I^W| = 0$, which is (0-1 1 1-1-1), then an entire SU_6 is left unbroken.

(4) If the little breaking is in the $\underline{27}$, then there are three candidate $|\Delta I^W| = 1/2$ weights. Each of these breaks $SU_2 \times U_1 \times U_1 \times U_1$ to U_1^{em}.

(5) As in an SU_3 model with Higgs scalars transforming as $\underline{3} + \underline{\bar{3}} + \underline{8}$, it often happens that scalars in different irreps get vacuum values and, for a range of parameters, their directions are perpendicular in weight space. The weak breaking in the standard SU_5 model transforms as $\underline{5} + \underline{\bar{5}}$, or the $\underline{10}$ of the SO_{10} theory, which suggests that the weak breaking L has weights (0 1-1 0 1 0) and (-1 0 1-1 0 0) of the $\underline{27}$. This leads to a nice breaking pattern. L is perpendicular to B if a = 2c and b = 3c, so that B breaks E_6 to $SU_2 \times SU_2 \times U_1 \times U_1 \times SU_3^c$, and L breaks this on down to $U_1^{em} \times SU_3^c$. See Table 1.

3. AN E_6 MODEL WITH COMPOSITE MUON AND TAU FAMILIES

There is a widespread belief that the standard model of electromagnetic, weak, and strong interactions correctly describes low energy data. The phenomenological success of the $SU_2 \times U_1 \times SU_3^c$ model with left-handed doublet and right-handed singlet fermions is offset by the necessity of determining a large number of parameters and assigning a large number of elementary fermions from the analysis of a huge amount of experimental data. This difficulty is due partly to the semisimple group structure of the theory, but even more significantly for our considerations here, it is also due to the large number of known quarks and leptons.

The problems associated with the semisimple group structure may be overcome by embedding the standard model into a unifying group, which we denote by G'. The smallest candidate for G' is SU_5.[9] In this model the large number of known quarks and leptons are assigned to a highly reducible representation, and the fermions arranged in this way still appears disorderly. There have been several suggestions for tidying up the situation: (1) perhaps there is a local or other kind of family symmetry that is embedded together with SU_5 and possibly other factors into a yet larger unifying algebra; (2) perhaps, following the hints of supergravity, the fermions belong to a large irrep of a relatively small group: (3) perhaps none of the known quarks and leptons are elementary; or (4) perhaps some of the known fermions are elementary, (for example, one family), and the rest are composite.

We explore the last alternative here.[2] The proposal is not very radical and at first glance, it would not seem possible for that kind of model to give a satisfying account of the proliferation of elementary fermions. Nevertheless, there is an attractive example where the electron family (or muon family) is elementary and the muon (or electron family) and the tau family are composites. The binding force, color and flavor are unified into the exceptional group E_6[4] and the elementary fermions are assigned to the $\underline{27}$.

We now "derive" this E_6 model, because such a discussion shows that the model is quite unique. Suppose that G' is a simple Lie group that unifies color and flavor in the usual way,[5] and that there is a U_1 factor U_1^t, not in G', with a current that is coupled to a vector boson that provides the binding force.[11] We assume that $G' \times U_1^t$ is a maximal subgroup of a simple group G, so the elementary particle fields are assigned to irreps of G. The branching rules derived from the embedding,

$$G \supset G' \times U_1^t \quad , \tag{22}$$

provide (up to an overall scale) the binding charge eigenvalue of each irrep of G'.

The second assumption is the existence of a short-range attractive force between two particles with Q^t charges of opposite sign. The composite fermions are s-wave bound states of the elementary fermions and certain scalar particles in the theory. There are vector bosons with nonzero Q^t, but we assume that they get super heavy masses, and so are irrelevant for this discussion. An advantage of the fermion-scalar binding picture is the ease with which the fermion helicity structure is maintained. In addition to the composite fermions, there may also be composite scalars that are fermion-antifermion bound states.

The next assumptions are phenomenologically motivated. For parity to be conserved in the electromagnetic and strong interactions, the fermion assignments including the composite ones, must be vectorlike under color and electric charge. With regard to the weak interactions, we assume that parity is violated because the theory is flavor chiral, so the left-handed fermions are in a complex representation. In addition, we assume that there is one family of elementary fermions.

We select the G' that is as small as possible. G' must have complex irreps, so the smallest candidate for G' is SU_5, with the elementary fermions transforming as $\underline{\bar{5}} + \underline{10}$. Now, $SU_5 \times U_1^t$ is a maximal subgroup of SU_6, Sp_{10}, and SO_{10}. There are no irreps of dimension less than 5000 of SU_6 or Sp_{10} that contain a $\underline{\bar{5}} + \underline{10}$ of

SU_5. SO_{10} is unacceptable because parity must be conserved in QED and QCD. The $\underline{\bar{5}}$ and $\underline{10}$ have different Q^t values, so the e^- and e^+ will bind to different scalars; similarly for d and $\bar{d}$.

We can avoid that difficulty by requiring that a left-handed fermion always be assigned to the same irrep of G' as its left-handed antiparticle image. This first happens for $SO_{10} \times U_1{}^t$, which is a maximal subgroup of SO_{12} and E_6. We discard SO_{12} because it gives a vectorlike theory for all interactions, even for the composite fermions. This leaves us with E_6 and the embedding,

$$E_6 \supset SO_{10} \times U_1{}^t \quad . \tag{23}$$

Table 3. Physical Roots and Axes in E_6 Weight Space.

	Dynkin Basis	Dual Basis
	(0 0 0 0 0 1)	[1 2 3 2 1 2]
Color Roots	(0 1 0 0 -1 0)	[1 2 2 1 0 1]
	(0 -1 0 0 1 1)	[0 0 1 1 1 1]
Weak Isospin Root	(1 0 0 0 1 -1)	[1 1 1 1 1 0]
Q^{em} axis	$\frac{1}{3}$(3 -2 3 -3 2 -2)	$\frac{1}{3}$[2 1 2 0 1 0]
$I_3{}^W$ axis	$\frac{1}{2}$(1 0 0 0 1 -1)	$\frac{1}{2}$[1 1 1 1 1 0]
Y^W axis	$\frac{1}{3}$(3 -4 6 -6 1 -1)	$\frac{1}{3}$[1 -1 1 -3 -1 0]
Q^t axis	3(1 -1 0 1 -1 0)	[1 -1 0 1 -1 0]
Q^r axis	(-3 -1 4 1 -1 -4)	[-1 1 4 3 1 0]

B axis (-3c, 3c-a, 2a-b-c, 2b-a-c, 2c-b, -a)

[-c c a b c 0]

L axis $(-3d\mp1, 2d-e\pm2, d+e\mp1, e-2d\mp1, 2d-e\pm2, d+e)$

$[-d, d\pm1, d+e, e, d\pm1, 0]$

The fundamental fields include the $\underline{78}$ vector bosons, the $\underline{27}$ of elementary fermions, and a $\underline{78}$ of scalars that can form bound states. The $\underline{78}$ scalars alone have no Yukawa couplings with the $\underline{27}$ of fermions, so there is at least an $E_6 \times E_6$ chiral invariance that may be broken dynamically by scalar bound states. This is a rather nice scenario, because, as we shall discuss, there is no gauge hierarchy problem.[12]

The eigenvalues of Q^t were computed in the last section. We list here the branching relations with the Q^t values in parenthesis:

$$\underline{27} = \underline{1}(4) + \underline{10}(-2) + \underline{16}(1) \tag{24}$$

$$\underline{78} = \underline{1}(0) + \underline{45}(0) + \underline{16}(-3) + \underline{\overline{16}}(3) \tag{25}$$

The normalization convention is set in Table 3.

We now use (24) and (25) to construct the bound state spectrum. We assume that neither the $\underline{27}$ nor $\underline{78}$ get superlarge masses, but are heavy enough and the binding strong enough that the composites appear pointlike to all probes made so far, such as in lepton pair production or g-2 experiments. The composite fermions are due to the binding of the $\underline{16}$ and $\underline{\overline{16}}$ in the $\underline{78}$ to the elementary fermions with Q^t of opposite sign. The spectrum of left-handed fermions, classified by SO_{10} irreps is

$$f_L = (\underline{1} + \underline{10} + \underline{16}) + (\underline{16}) + (\underline{16} + \underline{\overline{144}}) \quad . \tag{26}$$

The first set in (26) are the elementary fermions in the $\underline{27}$; the second set arises from the binding of the SO_{10} singlet fermion to the $\underline{16}$ of scalars, and is the most tightly bound; the third set arises from the binding of the $\underline{10}$ of fermions to the $\underline{\overline{16}}$ of scalars; and a fourth set that is omitted from (26) would be due to the binding of $\underline{16}$ and $\underline{16}$, which would give a set with $\underline{10}$ + $\underline{120}$ + $\underline{126}$. This last set is least strongly bound, if bound at all, and we shall neglect it for the remainder of the talk. A similar discussion of the scalar bound states in $\underline{27} \times \underline{27}$ is also possible, and these might be used to break the weak interaction group dynamically. The composite weights in the $\overline{\underline{27}}$ coincide with the example at the end of Sec. 2.

Perhaps the most amusing feature of (26) is the occurrence of three families of $\underline{16}$'s. The electron and muon families can be assigned to the first two $\underline{16}$'s; without further analysis it is not possible to decide which is elementary. The τ family should be either in the third $\underline{16}$ or in the $\overline{\underline{144}}$, which also contains a $\overline{\underline{5}} + \underline{10}$ of SU_5. Since the $\underline{16}$ and $\overline{\underline{144}}$ are bound with the same overall binding strength, a calculation is needed to decide which has the lowest mass states. What is more significant is that this model predicts a great proliferation of quarks, leptons, and other fermions not too far above the τ and b masses. Although this is not a unique prediction of this bound state model, it will be interesting to see what will be discovered above present PETRA energies.

For the composite fermion model to imitate physical reality, the binding force must be very strong and very short ranged. Thus the vector boson mediating the force must be very heavy (perhaps around 10 TeV) and the coupling $\alpha_t(Q^2)$ must be very large at small Q^2. Without adequate field theory technology to compute large running couplings, it is not possible to give quantitative results about the masses and the behavior of the binding force. However, the behavior of small couplings is well understood from the perturbation theoretic treatment of the renormalization group equations; since we know that the values of the standard model gauge couplings are small at Q^2 around 10 GeV^2, it is necessary to check whether it is indeed possible to compute $\alpha_t(Q^2)$ for all Q^2.

The argument that the gauge couplings do become large is one of self consistency. If the $\underline{78}$'s of vector bosons and of scalars and the $\underline{27}$ of fermions are the only contributions to the one loop approximation to the running coupling constant equations, then the theory is asymptotically free. (We ignore the scalar self couplings in this consideration.) Thus, if the strong coupling is small at 10 GeV^2, then it is even smaller at larger Q^2, and there is no reason to believe that any of the E_6 couplings get large. However, if the composite states are tightly enough bound that they also contribute to the one loop approximation over, say, the range of 100 GeV to 10 TeV, then they can destroy the asymptotic freedom of QCD and can push the coupling up into a region where we cannot, at present, compute it.

There are 41 effective flavors of quarks in (26), and in the one-loop approximation, we can estimate the value of $\bar{Q}^2$, where the QCD coupling becomes large. It is

$$\sqrt{\bar{Q}^2} = \sqrt{Q^2}\ \exp\left[-\left(\frac{6\pi}{\alpha_c(Q^2)}\right)\left(\frac{1}{33-2n_3-10n_6-12n_8}\right)\right] = \sqrt{Q^2}\ \exp[0.38/\alpha_c(Q^2)] \tag{27}$$

where "33" is the gluon contribution, n_3 is the number of Dirac quarks, n_6 is the number of Dirac $\underline{6}^c$, and n_8 is the number of Dirac octets. This rather rapid growth makes it possible to speculate that in a region between the weak boson mass and the unification mass, the gauge couplings may be large and $\alpha_t(Q^2)$ "freezes out" with a large value. We could also speculate that the couplings do eventually become small and asymptotically free before unification, (at large Q^2 the bound states no longer contribute) so that the proton decay rate is not too fast and the unification mass is not too close to the Planck mass. The dynamical symmetry breaking may cause the desert to bloom.

If this scenario is correct, then we must change attitudes toward some computations in unified models. Problems and advantages seem to be reversed over the situation with the usual SU_5 model. The unification mass, weak mixing angle, and quark masses are not easily computed, because the perturbation theory formulas do not hold over the whole extrapolation range. Thus there is no calculation of the proton lifetime, although this is not a difficulty of principle. The gauge hierarchy problem takes on a new character in the composite fermion model. Suppose the superstrong breaking is due to explicit Higgsism. Then at the unification scale, the composite scalars that do the weak breaking do not exist. The composite scalars appear elementary only on the scale of 100 GeV, and then may be available to do the weak breaking. Thus, the hierarchy problem would be resolved if the weak breaking were due to this dynamical mechanism. Further details of this model can be found in Ref. 2.

ACKNOWLEDGMENTS

I wish to thank Gordon Shaw for a very pleasant collaboration and Pierre Ramond for many helpful conversations.

REFERENCES

1. E.B. Dynkin, Amer. Math. Soc. Trans. Ser. 2, 6, 111 and 245 (1957).
2. G.L. Shaw and R. Slansky, Los Alamos Preprint (1980).
3. B.G. Wybourne, "Classical Groups for Physicists", Wiley-Interscience, New York (1974).
4. F. Gürsey, P. Ramond, and P. Sikivie, Phys. Lett. B60, 177 (1975).
5. M. Gell-Mann, P. Ramond, and R. Slansky, Rev. Mod. Phys. 50, 721 (1978).
6. W. McKay, J. Patera, and D. Sankoff, "Computers in Nonassociative Rings and Algebras", ed. R. Beck and B. Kolman, Academic Press, New York (1977), p. 235.
7. W. McKay and J. Patera, "Tables of Dimensions, Second and Fourth Indices and Branching Rules of Simple Lie Algebras", M. Dekker Inc., New York (1980); J. Patera and D. Sankoff, "Tables of Branching Rules for Representations of Simple Lie

Algebras", L'Université de Montréal, Montréal (1973).

8. R. Slansky, in preparation; P. Ramond, Caltech preprint CALT-68-945 (1980).

9. H. Georgi and S.L. Glashow, Phys. Rev. Lett. 32, 438 (1974), and H. Georgi, H. Quinn, and S. Weinberg, Phys. Rev. Lett. 33, 451 (1974).

10. J. Harvey, Caltech preprint CALT-68-737 (1979).

11. G.L. Shaw, Phys. Lett. 81B, 343 (1979).

12. E. Gildener, Phys. Rev. D14, 1667 (1976), and S. Weinberg, Phys. Lett. 82B, 387 (1979).

SO_{10} AS A VIABLE UNIFICATION GROUP

P. Ramond

California Institute of Technology

Pasadena, California 91125

The simplest unification of electroweak and strong interactions is realized by the SU_5 model of Georgi and Glashow[1] which incorporates the electric charge. There are, however, several indications that the SU_5 model is only a step in the right direction and not the definitive model, even though it gives a roughly successful account of the value for $\sin^2\theta_w$ and of certain mass ratios[2]. For one, there just are too many known elementary fermions, and in SU_5 left- and right-handed components of the same mass fermion must appear in different representations: each family of "comparable" mass is described by a reducible combination $\bar{5} + 10$ of left-handed Weyl spinors. The lightest such family contains e^-, ν_{eL}, $\vec{u}$ and $\vec{d}$, the additional families for the $\bar{\mu}$ and $\bar{\tau}$ are just clones of the first family[F1].

In the conventional model, the fermion masses are generated by terms of the form $\bar{5}_f \cdot 10_f$ and $(10_f \cdot 10_f)_s$ only, thus excluding

---

F1 Assuming, of course, that the t-quark is the missing component of the third τ-family. Perfectly viable models exists, such as the E_6 model of Gursey, Sikivie and the author where the t-quark does not exist. A likely experimental signal for such "topless" models will be the anomalous neutral current couplings of the τ-lepton which is vectorlike in a two-family E_6 model. However this type of model does not explain the multiplicity of fermion families and is therefore not entirely satisfactory.

$(\bar{5}_f \cdot \bar{5}_f)$ which gives rise to neutrino Majorana masses[F2]. Given the SU_5 group theory

$$\bar{\underset{\sim}{5}} \cdot \underset{\sim}{10} = \underset{\sim}{5} + \underset{\sim}{45} \quad ; \quad \underset{\sim}{10} \cdot \underset{\sim}{10} = (\bar{5} + \overline{50})_S + \overline{45}_A \quad ,$$

we see that Higgs field transforming as $\underset{\sim}{5}$, $\underset{\sim}{45}$ or $\underset{\sim}{50}$ can appear in the Yukawa couplings; of those, 50_H has no color and charge singlet component so that only the 5_H and 45_H can give fermions masses. Furthermore, in this form (i.e., no $\bar{5}_f \cdot \bar{5}_f$ fermion masses) the model after spontaneous breaking conserves baryon minus lepton number of the fermions, thus forbidding lepton number breaking neutrino Majorana masses.

The breaking of SU_5 into $SU_3^c \times SU_2 \times U_1$ is achieved by means of a $\underset{\sim}{24}$ of Higgs at a scale $\sim 10^{15}$ GeV[3], while in its simplest form the $SU_2 \times U_1$ breaking is done by a $\underset{\sim}{5}$ of Higgs. It is amusing to note that in this form ($\underset{\sim}{5}$ + $\underset{\sim}{24}$ of Higgs) the fermions acquire a calculable $\underset{\sim}{45}$ contribution to their masses coming from nonrenormalizable terms of the form $\bar{5}_f\ 10_f\ 24_H\ \bar{5}_H$ and $\bar{5}_f\ 10_f\ \bar{5}_H\ (24_H)^2$ which comes up at the two-loop level[4] and is therefore small (although it may be used to give mass to the light family). In order to explain the mass ratio $\frac{m_e}{m_\mu} \simeq \frac{1}{10}\frac{m_d}{m_s}$, it seems necessary to add some $\underset{\sim}{45}$ breaking as well as three different $\underset{\sim}{5}$'s[F3], as noted by Georgi and Jarlskog[5].

F2 Majorana neutrino masses ($\Delta I_W = 1$) can be artificially added in the Weinberg-Salam model by adding a complex triplet of Higgs.

F3 If one requires no terms of dimension two or three to ensure naturalness of the Georgi-Jarlskog scheme, one has to gauge an anomaly-free "family" U_1 to avoid a massless Goldstone boson, T.J. Goldman, P. Ramond, D.A. Ross, unpublished. Note that this scheme can be made natural by introducing only two 5's and a 45 of Higgs, but then terms of dimension-two are needed to ensure naturalness without massless bosons. P.Ramond, unpublished. See also S. Nandi and K. Tanaka, Ohio University preprint COO-1545-270, (1980. This latter scheme does not generalize to SO_{10}.

Another exciting feature of unified theories is their potential for explaining the observed baryon asymmetry of the universe starting from symmetric initial conditions, by means of CP- and baryon number violating processes at high temperature[6]. However the SU_5 model is not entirely satisfactory in this respect since in it soft CP-violation is linked to electroweak breaking which is presumably a low temperature effect[F4], unless CP-violation is put in ab initio by making certain coupling constants complex. Also the picture of a single breaking seems to run in trouble with the observed lack of monopoles in our vicinity[7].

The next step in unification is at the level of the SO_{10} model[8]. There each family is represented by an irreducible complex spinor representation with an SU_5 content

$$16_f = \bar{5}_f + 10_f + 1_f \quad ,$$

which includes an additional neutral left-handed spinor. This larger unification still does not explain the family cloning, and should be regarded only as a further step in the right direction[F5]. Still it has several noteworthy features which we want to investigate.

When $SU_2 \times U_1$ is broken down the extra neutral lepton in each 16_f becomes the right-handed partner to the regular left-handed neutrinos thus giving neutrinos unacceptably large masses of the

F4 High temperatures and high Q^2 are by no means synonymous. In fact with a complicated Higgs structure it is possible to appeal to the electroweak breaking. See R. Mohapatra and Senjanovic, CCNY preprint 1979.

F5 In this respect, one is led to consider large complex spinor representations of SO_{4N+2} groups. An example is a model based on the $\underset{\sim}{1024}$ spinor representation of SO_{22}, M. Gell-Mann, P. Ramond, R. Slansky in the Proceeding of the Stony Brook Conference on Supergravity, September 1979.

order of the charge 2/3 quark masses. The fermion mass matrix is ruled by SO_{10} groupology

$$\underset{\sim}{16} \times \underset{\sim}{16} = (\underset{\sim}{10} + \underset{\sim}{126})_S + (\underset{\sim}{120})_A \quad .$$

Gell-Mann, Slansky and the author[9] have suggested that the complex $\underset{\sim}{126}$ can be used to give the extra lepton a large Majorana SU_5-singlet mass. Then the neutral lepton mass matrix is per family of the form

$$\begin{pmatrix} 0 & a \\ a & A \end{pmatrix}$$

where, up to Yukawa to gauge coupling ratios, a is $\sim$300 GeV, and A is of the order of $10^{12} - 10^{15}$ GeV where SO_{10} is broken down to SU_5. This results in a small Majorana mass for the left-handed neutrino of order $\frac{a^2}{A}$. An appealing scenario then emerges: use a 45_H to break SO_{10} down to $SU_4 \times SU_2 \times U_1$ at say 10^{15} GeV, and then 126_H to break $SU_4 \times SU_2 \times U_1$ to $SU_3^c \times SU_2 \times U_1$ at $\sim 10^{12}$ GeV. Then, as noted by Georgi and Nanopoulos[10] the value of $\sin^2\theta_W$ is a bit larger than in SU_5 (a good sign).

Also, in our scheme the neutrinos can acquire masses $\sim$.1 eV, with the result of possibly observable neutrino oscillations. Still another bonus appears[11]: the 126_H acquires in general a complex vacuum expectation value and its phase cannot be rotated away because of terms like $(126_H)^4$ in the Higgs potential. Hence it provides a ready-made mechanism for time reversal violation at high temperature.

In passing let us note that the appearance of lepton number violating neutrino Majorana masses is symptomatic of the irreducibility of the fermion representation which tends not to allow global conservation laws to survive: baryon number at the level of SU_5 and lepton number at the level of our SO_{10} scheme.

We[11] now present a three-family SO_{10} model which incorporates our neutrino mass scheme, as well as the Georgi-Jarlskog mass relation, $\tan^2\theta_c \simeq \frac{m_d}{m_s}$, and no strong CP-violation. As might be expected, the Higgs structure is quite complicated but not without a certain systematic structure of its own. The Yukawa coupling is given by

$$(A16_1\cdot 16_2+B16_3\cdot 16_3)\overline{126}_1 + (a16_1\cdot 16_2+b16_3\cdot 16_3)(10_1+i\ 10_2)$$

$$+ c16_2\cdot 16_2\ \overline{126}_2 + d16_2\cdot 16_3\ \overline{126}_3 + h.c.$$

where 16_i refers to the left-handed fermions, 10_1, 10_2, 126_i are the Higgs, and A,B,a,b,c,d are real coupling constants. The model is technically natural, and has no massless Goldstone bosons; they are avoided by terms of the form $(126_1)^4 + (\overline{126}_1)^2$, $(10_1+i\ 10_2)^4 + (10_1-i\ 10_2)^4$ and $(126_2)^4 + (\overline{126}_2)^4$ in the potential. In addition the model contains a 45 of Higgs which does not couple to fermions. No terms of dimension 2 or 3 are needed for naturalness reasons in the potential, making the model amenable to a Coleman-Weinberg treatment[12]. The SU_5 decomposition of the 126,

$$\underset{\sim}{126} = \underset{\sim}{1} + \underset{\sim}{\bar{5}} + \underset{\sim}{10} + \underset{\sim}{\overline{15}} + \underset{\sim}{\overline{45}} + \underset{\sim}{\overline{50}} \quad ,$$

shows that possible v.e.v. lie along $\underset{\sim}{1}$, $\underset{\sim}{\bar{5}}$, $\underset{\sim}{45}$ or $\underset{\sim}{\overline{15}}$ [F6].

F6 The $\underset{\sim}{126}$ is a complex representation even though it is built out of a totally antisymmetrized fifth rank tensor. Call α a group of five SO_{10} indices totally antisymmetrized. Then given a 252-dimensional tensor T_α, we can project out the 126 and $\overline{126}$ by means of the Levi-Civita symbol $\varepsilon_{\alpha\beta} = -\varepsilon_{\beta\alpha}$ as follows:

$$126 \sim (\delta_{\alpha\beta} + \varepsilon_{\alpha\beta})T_\beta \qquad\qquad \overline{126} \sim (\beta_{\alpha\beta} - \varepsilon_{\alpha\beta})T_\beta \quad .$$

Thus $i\varepsilon_{\alpha\beta}$ stands as the charge conjugation operator that transforms 126 into $\overline{126}$.

We assume that the Higgs potential is such that we can choose in SU_5 space

$$\langle 126_1 \rangle \sim \underset{\sim}{1} \text{ of } SU_5$$

with a scale of order $10^{12} - 10^{15}$ GeV; $\langle 126_2 \rangle \sim \underset{\sim}{45}$, $\langle 126_3 \rangle \sim \bar{5}$[F7], $\langle 10_1 \rangle$, $\langle 10_2 \rangle$ all at a scale of order 300 GeV. Note that in terms of SU_5,

$$\underset{\sim}{10} = \underset{\sim}{\bar{5}} + \underset{\sim}{5} \quad .$$

The large $\langle 126_1 \rangle$, together with the large $\langle 45 \rangle$ are sufficient to break SO_{10} to $SU_3^c \times SU_2 \times U_1$. The mass matrices in the e^-, μ^-, τ^- family basis are given by

$$\begin{pmatrix} 0 & Re^{i\theta} & 0 \\ Re^{i\theta} & Se^{i\chi} & 0 \\ 0 & 0 & Te^{i\theta} \end{pmatrix} \quad \text{for charge } -1/3 \text{ quarks} \quad ,$$

$$\begin{pmatrix} 0 & Pe^{i\delta} & 0 \\ Pe^{i\delta} & 0 & Qe^{i\mu} \\ 0 & Qe^{i\mu} & Ue^{i\delta} \end{pmatrix} \quad \text{for charge } 2/3 \text{ quarks} \quad ,$$

$$\begin{pmatrix} 0 & Re^{i\theta} & 0 \\ Re^{i\theta} & -3Se^{i\chi} & 0 \\ 0 & 0 & Te^{i\theta} \end{pmatrix} \quad \text{for charge } -1 \text{ leptons} \quad .$$

Interestingly, the phase of the determinant of the quark mass matrix depends only on the angle between the $\langle 10_1 \rangle$ and $\langle 10_2 \rangle$ values; in

F7 It is tempting to believe that this orthogonality of vacuum expectation values in SU_5 space points to certain, as yet unknown, systematics in the Higgs system.

particular if they are orthogonal ($\langle 10_1\rangle \cdot \langle 10_2\rangle = 10$), the angle vanishes - this can be arranged by choosing the sign of the $(10_1 + i\ 10_2)^4 + (10_1 - i\ 10_2)^4$ term in the potential. Assuming that the calculable deviation from zero for θ_{QFD} is small, this avoids a possibly large value for the neutron electric dipole moment[13].

The neutral lepton mass matrix is 6 × 6 dimensional and not worth jotting down here. Suffice it to say that all neutrinos are massive and that its diagonalization leads to flavor neutrino mixing of the order of θ_c.

The $\langle 126_1\rangle$ value can be made complex by fixing the sign of the $(126_1)^4 + (\overline{126}_1)^4$ term in the potential, thus leading to time reversal violation at high temperature. Note also that a "low" value for $\langle 126_1\rangle$ might go in the right direction to explain the large value of $\sin^2\theta_W$ (w.r.t. SU_5), lack of monopoles and detectable neutrino oscillations. Details of this model will be presented elsewhere[14].

To conclude, let us emphasize that models with complicated Higgs structure need not alarm the reader. It is likely that some of these can be generated dynamically. There are still many mysteries to be unravelled, such as the ratio between unification and electroweak breaking, the role of quantum gravity, etc.

It is a pleasure to thank my collaborators J. A. Harvey and D. B. Reiss for many useful discussions and comments.

REFERENCES

1. H. Georgi and S. L. Glashow, Phys. Rev. Letters 32, 438 (1976).
2. A. Buras, J. Ellis, M. Gaillard, D. Nanopoulos, Nucl. Phys. B135 66 (1978).
3. T. J. Goldman, D. A. Ross, Phys. Letters 84B, 208 (1979); W. J. Marciano, Phys. Rev. D20, 274 (1979).
4. J. A. Harvey, P. Ramond, D. B. Reiss, unpublished.
5. H. Georgi and C. Jarlskog, Harvard preprint HUTP/79/A026.
6. See for instance, D. Nanopoulos and S. Weinberg, Harvard pre-

print HUTP/79/A026 which contains numerous references.

7. J. P. Preskill, Phys. Rev. Letters 43, 1365 (1979).
8. H. Georgi in Particles and Fields 1975, (AIP Press, New York); H. Fritzsch, P. Minkowski, Ann. Phys. 93, 193 (1975).
9. M. Gell-Mann, P. Ramond, R. Slansky unpublished, and P. Ramond, Sanibel Symposia Talk, February 1979, CALT 68-709.
10. H. Georgi and D. Nanopoulos, Harvard preprint HUTP/79/A039.
11. J. A. Harvey, P. Ramond, D. B. Reiss, Caltech preprint CALT 68-758.
12. S. Coleman, E. Weinberg, Phys. Rev. D7, 1888 (1973).
13. M. A. B. Bég, H. S. Tsao, Phys. Rev. Letters 41, 278 (1978); R. N. Mohapatra, D. Wyler, SLAC preprint SLAC-PUB-2382.
14. J. A. Harvey, P. Ramond, D. B. Reiss, in preparation.

MIGDALISM REVISITED: CALCULATING THE BOUND STATES OF QUANTUM CHROMODYNAMICS*

Paul M. Fishbane

University of Virginia

Charlottesville, Virginia 22901

ABSTRACT

We discuss a scheme to determine the location of bound states in QCD using calculable functions in the asymptotic regime and the assumption of confinement. We discuss tests of this idea, an example worked to fourth order, and ways to carry the ideas further.

In this paper I review an idea first proposed[1] by A.A. Migdal in which the assumption of confinement in QCD coupled with the calculation of two-point functions in an asymptotic region, where perturbative results apply, leads to a determination of the location of the bound states of the theory as well as their residues. This idea has been mutated, refined, and tested by Migdal[2], as well as by P. Kaus[3,4,5] and his collaborators, and has been the object of recent attention by S. Gasiorowicz, P. Kaus, and the author.

I have divided this talk into four sections. Section I contains the formalism which, given a quantity such as a two-point function (which contains the bound state poles of the theory) in an asymptotic

*Work supported in part by the National Science Foundation under grant no. NSF-PHY-79-01757.

region, leads to a unique determination of the poles and their residues. In Section II the form and properties of the two-point functions themselves are discussed. Section III contains discussion of sample calculations both of potential theory and of QCD to order α^2, and the last part contains a discussion.

I. FINDING THE POLES

We start with some 2-pt. Green's function $G_{PT}(x)$ at small x, labelled with its subscript to indicate that perturbation theory is involved in the calculation,

$$G_{PT}(x) = \langle 0|T\Big(O^{(1)}(x)\ O^{(2)}(0)\Big)|0\rangle \quad .$$

The $O^{(i)}$ are some appropriate operators formed from quark and gluon fields whose quantum numbers determine the quantum numbers of the bound state poles to be determined. In momentum space $G_{PT}(t)$ will have the large t behavior

$$G_{PT}(t) = At^{\nu}\ (1 + \text{non-leading}) \quad . \tag{1.1}$$

Let us now suppose that QCD exhibits confinement. This means there are no cuts associated with a continuum of gluons and quarks (There may of course be cuts associated with the physical decay of hadrons into other hadrons. These can be suppressed[6] in the theory by working in the limit of the number of colors $N_C\to\infty$. Physical thresholds would appear as N_C^{-1} corrections and could thus be controlled.) G_{PT} has a cut structure in general and could therefore not be the true Green's function of the theory, G(t). Our aim is to construct a cut-less G(t), closer to $G_{PT}(t)$ than any power in the asymptotic region. We shall see that when this is done, G is determined up to a CDD ambiguity, including its poles and residues.

To remove the cut from G_{PT}, we write

$$G = G_{PT} + N/D \quad , \tag{1.2}$$

where D is an entire function whose zeros are the poles of G. N is given by

$$N(t) = \frac{1}{\pi}\int_0^\infty \frac{\mathrm{Im}(G(t')-G_{PT}(t'))}{t'-t} D(t')dt' = \frac{1}{\pi}\int^\infty \frac{\mathrm{Im}G_{PT}(t')}{t'-t} D(t')dt' \quad . \tag{1.3}$$

Thus

$$G = G_{PT} - \frac{1}{\pi D}\int \frac{\mathrm{Im}\ G_{PT}}{t'-t} D\ dt' \quad . \tag{1.4}$$

G clearly has no branch points. In addition D is determined by the requirement that $G - G_{PT} \to 0$ faster than any power of t as $|t|\to\infty$ and that G has poles with positive residues. Namely, if D increases faster than any power, then D has infinitely many zeros. (This argument is thus equivalent to the expectation that a confined system will have an infinite number of poles, i.e. D will have an infinite number of zeros.) We now couple this remark with the requirement of positive residues, which states that the zeros of N must interlace those of D, so that the residue of N/D can be positive at the zeros of D. N itself must then have an infinite number of zeros and will decrease faster than any power. On the other hand, Eq. (1.3) can be expanded as a power series in $(1/t)^n$; the coefficients of these powers must thus vanish, which means

$$\int_0^\infty t'^p\ \mathrm{Im}\ G_{PT}(t')\ D(t')\ dt' = 0\ , \quad p \geq 0 \quad . \tag{1.5}$$

Eq. (1.5) is a set of moment conditions whose solution will determine D.

We now turn to the problem of solving the moment conditions (1.5). We organize the terms in G_{PT} by writing

$$\mathrm{Im}\ G_{PT} = t^\nu\big(1 + \lambda f_1(t) + \ldots + \lambda^n\ f_n(t) + \ldots\big)\ ,$$

where, e.g. $f_j(t)$ is a decreasing power t^{-j}. (Logarithms can easily by a derivative of a power, $\lim_{\varepsilon\to 0} \frac{d}{e\varepsilon} t^{\varepsilon}$.)

Then set

$$D = D_o(t) + \ldots + \lambda^n D_n(t) + \ldots \quad .$$

By grouping powers of λ we pick out the $0(\lambda^n)$ term in (1.5),

$$\int_0^\infty dt\, t^{\nu+p} \left[D_n(t) + \sum_{m=o}^{n-1} f_{n-m}(t)\, D_m(t) \right] = 0, \quad p \geq 0 \quad . \tag{1.6}$$

These equations are handled consecutively, starting with $0(\lambda^o)$,

$$\int_0^\infty dt\, t^{\nu+p}\, D_o(t) = 0 \quad , \quad p \geq 0 \quad .$$

The solution to this equation is

$$D_o(t) = \sum_{m=o}^{M} b_m \, (R\sqrt{t})^{-\nu+m} \, J_{\nu+m-M}\,(R\sqrt{t}) \quad ,$$

where R, M, and the b_m are all arbitrary. Clearly the zeros of this function are not uniquely determined unless the b_m are given, which is equivalent to information about the asymptotic behavior of $D_o(t)$. To eliminate this ambiguity we make an assumption of minimal dynamics, in which all b_m except b_o vanish; a last overall constant b_0 plays no role in the location of zeros. This assumption is closely related to the existence of CDD poles, in which bound states are elementary particles rather than dynamical ones, and so is very much within the spirit of determining bound states in terms of the quark-gluon dynamics. In sum, then,

$$D_o(t) = b_o \, (R\sqrt{t})^{-\nu} \, J_\nu(R\sqrt{t}) \quad . \tag{1.7}$$

The function on the right is a "cut-less Bessel function." The asymptotic behavior t^{ν} thus gives poles at the zeros of the Bessel function $J_{\nu}(R\sqrt{t})$.

Having solved the λ^{0} equation, the remaining equations are consecutively solved by a Green's function.[7] Thus to $O(\lambda^{1})$,

$$\int dt\ t^{\nu+p}\ D_1 = -\int dt\ t^{\nu+p}\ f_1(t)\ D_0(t)$$

is, for f_1 a power of t, solved[7] by the Green's function

$$G_{\nu}(t,t') = -\frac{1}{2}\sum_{p=1}^{\infty} J_{\nu+p-1}(R\sqrt{t'})\ J_{\nu+p}(R\sqrt{t})\ \frac{(R\sqrt{t'})^{\nu+p-1}}{(R\sqrt{t})^{\nu+p}}\ , \qquad (1.8)$$

so that

$$D_1(t) = \int dt'\ G_{\nu}(t,t')\ f_1(t')\ D_0(t')\ .$$

All integrals can be done compactly. Alternatively, a simple ansatz[5] for the solution when the $f_i(t)$ are decreasing powers can be systematically handled.

We conclude this Section with three remarks:

(a) Given D(t) it is possible[5] to systematically construct an "equivalent potential," which plays the role of a potential in a Schrödinger equation. This equivalent potential is the real potential in tests of the method based on potential theory. Thus the construction unambiguously leads to the concept and form of a confining potential.

(b) The scale R is a free parameter. How it is treated is a matter of choice. One could fit one pole with this parameter. Alternatively another method[2] exists, the "α-expansion," which determines R in some optimum way based on the position of all the poles.

(c) The bound states best determined by this procedure are the lowest-lying ones. Corrections to G_{PT} give corrections to the higher-lying poles. This is simply because the large t behavior probes the short range part of the potential, and this is the part of the potential which describes the lowest-lying bound states.

II. GETTING THE TWO-POINT FUNCTIONS

We begin by discussing the proper analogue to the quantity G_{PT} in potential theory. This will allow us (Section III) to test the concept and method discussed here for problems which are exactly soluble. We are looking for a function which contains the bound state poles of the problem: this is the S-matrix. However, for confining potentials there are no asymptotic states and hence no S-matrix in the usual sense. A suitable generalization is[3]

$$\hat{S}(\nu,k^2) \equiv f(-\nu,k^2)/f(\nu,k^2) \quad , \tag{2.1}$$

where $f(\nu,k^2)$ is the usual Jost function and $\nu = j + 1/2$.

For the relativistic case we take guidance on the appropriate local operators $0^{(i)}(x)$ from our prejudices about the relevant structure of hadrons. Thus let us consider $q\bar{q}$ meson states, for which a stringlike form might be indicated:

$$S_\alpha \equiv \bar{\psi}(x)\ \gamma_\alpha \left[P \exp ig\int_x^y B_\mu\ dx^\mu\right] \psi(y) \quad . \tag{2.2}$$

Here we have chosen γ_α just as an example; longer chains of γ_μ's as well as γ_5's are generally expected. P is an ordering operator for the color matrices subsumed in the field B_μ.

We cannot use S_μ as an operator in the 2-pt. function because it is non-local. However, we can find a set of local operators equivalent to S_μ by expanding (2.2) in $\delta = y-x$. Such an expansion gives

$$S_\alpha = \bar{\psi}(x)\,\gamma_\alpha\,\psi(x) + \delta^\mu\,\bar{\psi}(x)\,\gamma_\alpha\,\nabla_\mu\,\psi(x)$$

$$+ \frac{\delta^\mu\,\delta^\nu}{2!}\,\bar{\psi}(x)\,\gamma_\alpha\,\nabla_\mu\,\nabla_\nu\,\psi(x) + \ldots \quad ,$$

where

$$\nabla_\mu = \partial_\mu + ig\,B_\mu$$

is the gauge covariant derivative. Thus a set of local operators to study would be

$$O^{(i)}(x) = \bar{\psi}(x)\,\gamma_{\mu_1} \ldots \gamma_{\mu_n}\,\nabla_{\nu_1} \ldots \nabla_{\nu_m}\,\psi(x) \quad . \qquad (2.3)$$

We could include γ_5 operators here as well. A similar operation for the pure gauge field operator

$$\mathrm{Tr}\; P \exp ig \oint B_\mu\, dx^\mu$$

would also give local operators, e.g. $F_{\mu_1\nu_1}(x)\ldots F_{\mu_n\nu_n}(x)$.

We form our two-pt. functions from the operators (2.3)

$$G_{PT}(t) = \int dx\; e^{ik\cdot x} \langle 0|T\Big(O^{(i)}(x)\; O^{(j)}(o)\Big)|0\rangle \quad . \qquad (2.4)$$

The superscripts on the operators carry much information. Let us first consider the angular momentum J, in which the matrix element in (2.4) is diagonal. We imagine that $O^{(i)}$ has a definite spin J: it is symmetric and traceless in its (J) free Lorentz indices. In this case $\mathrm{Im}G_{PT}$ has the form

$$\mathcal{G}_{\mu_1..\mu_J,\nu_1..\nu_J}(t) = P_{\mu_1..\mu_J,\nu_1..\nu_J}\left(\frac{k}{|k|}\right) f^J(t) \quad . \qquad (2.5)$$

The spin tensor $P_{\mu_1 \cdot\cdot \mu_J, \nu_1 \cdot\cdot \nu_J}$ has the form

$$P_{\mu_1 \cdot\cdot \mu_J, \nu_1 \cdot\cdot \nu_J} = \frac{1}{(2J-1)!} \frac{\partial}{\partial p_{\mu_1}} \cdot\cdot \frac{\partial}{\partial p_{\mu_J}} \frac{\partial}{\partial p'_{\nu_1}} \cdot\cdot \frac{\partial}{\partial p'_{\nu_J}} \{ (p^2 - \frac{(p\cdot k)^2}{k^2})^{5/2} \times$$

$$\times P_J \left(\frac{p\cdot p' - \frac{p\cdot k \; p'\cdot k}{k^2}}{\sqrt{p^2 - \frac{(p\cdot k)^2}{k^2}} \sqrt{p'^2 - \frac{(p'\cdot k)^2}{k^2}}} \right)$$

$$(p'^2 - \frac{(p'\cdot k)^2}{k^2})^{J/2} \} \quad , \tag{2.6}$$

where P_J is a Legendre polynomial. For example, some special caases are

$$J = 1 : \qquad P_{\mu\nu} = g_{\mu\nu} - \frac{k_\mu k_\nu}{k^2}$$

$$J = 2 : \qquad P_{\mu_1\mu_2, \nu_1\nu_2} = [\frac{1}{2} P_{\mu_1\nu_2} P_{\nu_1\mu_2} + \frac{1}{2} P_{\mu_1\nu_1} P_{\mu_2\nu_2} - \frac{1}{3} P_{\mu_1\mu_2} P_{\nu_1\nu_2}] \quad .$$

The above (dimensionless) spin tensors do not enter into the pole determination, and Im G_{PT} (t) = f^J (t).

To take a simple example (to be treated in more detail below), the operator $\bar{\psi}(x)\gamma_\mu\psi(x)$ at each point of the 2-point function (see fig. 1) gives

$$\mathcal{G}_{\mu\nu} \sim (g_{\mu\nu} - \frac{k_\mu k_\nu}{k^2}) \cdot k^2 \quad . \tag{2.7}$$

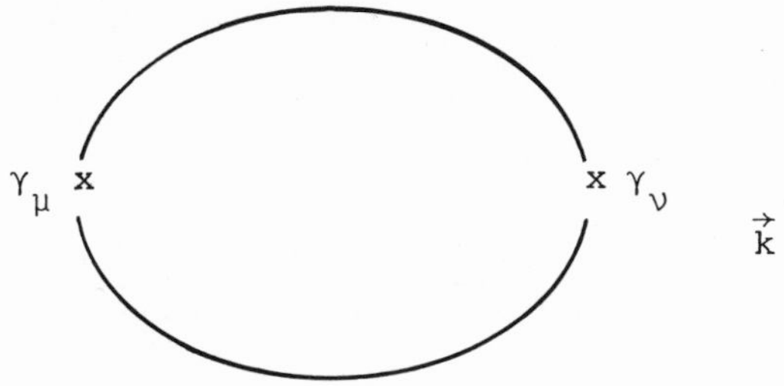

Fig. 1

According to the discussion above, the corresponding spin=1 bound states are at the zeros of J_1 $(R\sqrt{t})$. The left-most pole (see Fig. 2) we call the ρ, with others referred to as daughters of the ρ.

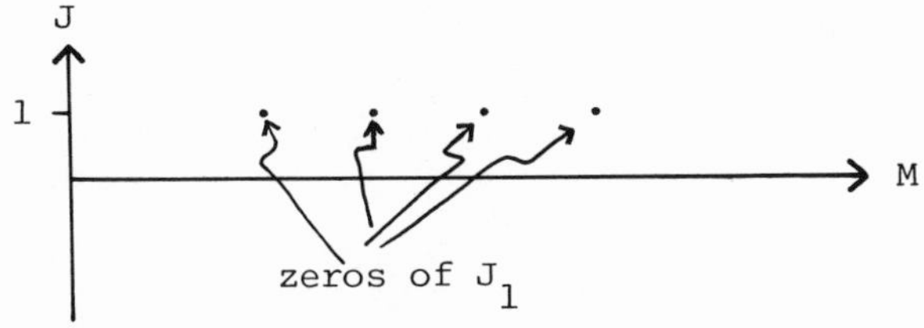

Fig. 2

Spin J recurrences come from higher twist (i.e. higher dimensionality) operators of the same spin. In the example above, such an operator is $\bar{\psi}(x)\gamma_\mu\nabla_\alpha\nabla^\alpha\psi(x)$. These operators lead to many additional states. Moreover the two point function is not in general diagonal in operators of the same spin but different twist, which further increases the count of states.

Dimensional analysis gives the zeroth order behavior of $f(k^2)$. Namely, if n is the number of free indices on G and τ is the number of contracted pairs, then $f^J(k^2) = (k^2)^\Delta$, where $\Delta = n + 2\tau$ is the dimensionality (note twist $= n+2+2\tau-J$). To $O(\alpha)$ this naive dimensionality is corrected by the anomalous dimension γ,

$$\Delta \rightarrow \Delta + N_c \frac{\alpha}{2\pi} \gamma \quad , \tag{2.8}$$

where

$$\gamma = \frac{1}{2} - \frac{1}{n(n+1)} + 2 \sum_{\ell=1}^{\infty} \frac{n-1}{(\ell+n)(\ell+1)} \quad . \tag{2.9}$$

III. EXAMPLES

We discuss here both potential theory examples which can be compared to exact results and a calculation of the spin one two-point function in QCD. In perturbation theory study has been made of the square well[3], the harmonic oscillator,[3,4] and[5] the potential $V = \lambda^2 \tanh^2(g^2 r/\lambda)$ (fig.3) This last potential, soluble[6] for $\ell = 0$, is a harmonic oscillator at the origin and has a threshold, i.e. is not confining. The general technique here is to use the exact solution to find the exact $\hat{S}(\nu,k^2)$, write the large k^2 limit of $\hat{S}$,

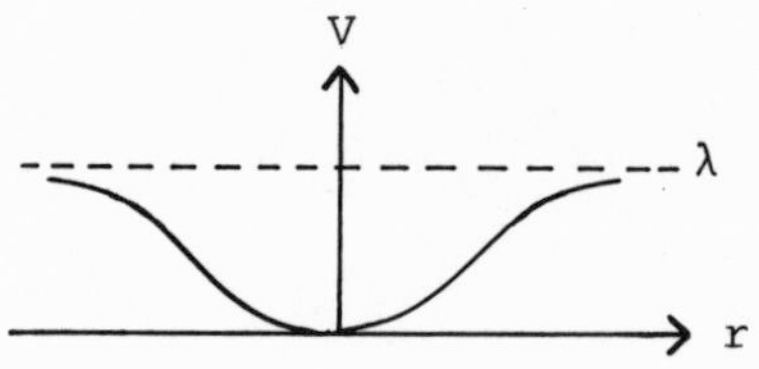

Fig. 3

$$\hat{S} \rightarrow A(-k^2)^\nu \left(1 + \frac{a}{k^2} + \frac{b}{k^4} + \ldots\right) \quad , \tag{3.1}$$

and then to use (3.1) by itself as the input for solution of the moment conditions, to test whether and with how many terms there is convergence to the exact bound state poles.

The square well is a trivial case, since its asymptotic $\hat{S}$ is a pure power, and its poles are in fact determined exactly by the zeros of Bessel functions. In this case R is just the size of the well. However, the method also works very well[3-5] for the other cases. The third potential in particular is interesting because it is not confining and signals this by an accumulation of poles near the threshold. The method is nevertheless able to account well for bound states lying within a radius corresponding to the k^2 for which the number of terms kept in $\hat{S}$, Eq. (3.1), is a good approximation to $\hat{S}$. These examples give some confidence that the method can be a practical one.

We next turn to an example to $O(\alpha^2)$ in QCD of the 2-point function diagonal in $\bar{\psi}(x)\gamma_\mu\psi(x)$. This corresponds to calculation of the photon self-energy function Π and has been calculated recently[9] because it gives higher order corrections to the important quantity R of e^+e^- annihilation. In our language we have

$$G_{PT} = q^2\,\Pi(q^2)$$

$$= \frac{1}{3} q^2 \,\ell n \frac{-q^2}{\mu^2} + \alpha \frac{C_F}{4\pi} q^2 \,\ell n \frac{-q^2}{\mu^2} + \frac{1}{\pi^2}\alpha^2 C_F\{-\frac{\beta_0}{32}\ell n^2(-\frac{q^2}{\mu^2}) \tag{3.2}$$

$$+ \ell n(-\frac{q^2}{\mu^2})(\frac{\beta_0}{16}(\ell n\, 4\pi - \gamma - 2M + 8D) - \frac{3B}{2})\}$$

or

$$\mathrm{Im}G_{PT} = \pi q^2 \quad \frac{1}{3} + \frac{\alpha}{4\pi} C_F + (\frac{\alpha}{4\pi})^2 C_F \beta_0 [-\ell n \frac{q^2}{\mu^2} + \ell n\, 4\pi - \gamma - 2M + 8D - \frac{24B}{\beta_0}] \} \, . \tag{3.3}$$

In this equation, N = number of colors and N_f = number of flavors, $C_F = (N^2-1)/2N$, $\beta_0 = 11/3N - 2/3N_f$, $D \approx .05$, $B \approx .02\, C_F - .05\, N + .006\, N_f$, and M is a renormalization scheme dependent constant, M = 0 in the scheme of ref. 10 and $\ell n 4\pi - \gamma$ in the scheme of ref. 11.

Note that since the $O(\alpha)$ term is a constant rather than $O(\ell n q^2)$, the anomalous dimension is zero for this term; this agrees with

Eq. (2.9), which gives $\gamma = 1$ for $n = 1$. However, since the $O(\alpha^2)$ term has a single power of $\ln q^2$, this piece plays the role of an anomalous dimension. This is transparent when we write

$$\text{Im } G_{PT} = \text{const. } (tR^2) \{ 1 + A \ln tR^2 \} \quad ,$$

where $A = O(\alpha^2)$, so that alternatively to this order

$$\text{Im } G_{PT} = \text{const. } (tR^2)^{1+A} \quad . \tag{3.4}$$

The zeros corresponding to bound state poles are then the zeros of $J_{1+A}(\sqrt{t}\, R)$.

Numerical discussion of these zeros will be presented elsewhere It is sufficient to state here that the trajectory corresponding to J_{1+A} is very close to J_1. We can thus conclude that order-by-order perturbative corrections seem to be a difficult way to improve the trajectory. We discuss this further in the next Section.

IV. CONCLUSION

Several remarks are in order on the results of the examples presented:

(i) The "correction", ~t log t of the perturbative calculation formally leads the first term, ~t. This correction is a correction in the sense of the leading logarithm calculation scheme which has been so successful in other contexts of QCD; namely, that $\alpha \log t$ should be regarded as small compared to one. This suggests that one possibility to avoid the slow convergence of term-by-term perturbative expansion is to use summed leading log results. For example[12], Cornwall and Tiktopoulos suggest that for pure glue the two point function will have a form like $\exp \oint dl_1 \oint dl_2\, \Delta(x-y)$, where l_1 and l_2 are loops in the neighborhood of x and of y, and where

$$\Delta(x-y) = \int d^4k\, e^{ik\cdot x}\, g^2(k)/k^2 \; ,$$

g(k)=running coupling constant.

The problems arising here are technical ones of the spin projection and the solution of the moment conditions.

(ii) For large values of the argument, the zeros of J_{1+A}, like those of J_1 and indeed any J, are spaced equally in the argument. Since this argument is $\sim\sqrt{t}$, we see that the asymptotic spacing of the bound states corresponds to square root, not linear, trajectories. (Although we discuss here the spacing in mass for a given angular momentum rather than the usual angular momentum trajectory, reflection shows these are equivalent.) One possible attitude to this is that only a few terms in the asymptotic behavior do not allow one to expect an accurate description of the high-lying poles. Another possible view is indicated by the fact that inverse power corrections for the harmonic oscillator lead to linear trajectories, so that what is required is

(iii) inverse power rather than logarithmic corrections. One (possibly trivial) source of such inverse powers is the quark mass m. In zeroth order, for the same two-point function discussed in Section III for example, we have

$$\begin{aligned} \mathrm{Im}\, G_{PT} &= \frac{\pi}{3}\, t\, (1 + \frac{2m^2}{t}) \sqrt{1-\frac{4m^2}{t}} \\ &= \frac{\pi}{3} (1 + \frac{4m^2}{t'})^{-1/2} (t' + 6m^2) \quad , \end{aligned} \tag{4.1}$$

where

$$t' = t - 4m^2 \quad .$$

We have written G_{PT} in terms of t' to emphasize that this is a more natural variable, and one which is more relevant to the observed families of hadrons. Eq. (4.1) has of course an infinite series of t^{-1} or t'^{-1} corrections. It has also been suggested[4] that instanton corrections[13] could provide other power corrections;

the scale here would not be the quark masses but the cutoff characteristic of the postulated confining phase change, i.e. a typical hadronic scale.

(iv) Finally we mention that it is nowhere engraved in stone that it is necessary to study the two-point function (although this is what survives when $N_C \to \infty$); rather any function that contains the bound states as poles will do in principle. This may allow us to take advantage of other summed forms of perturbation theory, e.g. the eikonal.

ACKNOWLEDGEMENTS

I report here on work done in collaboration with S. Gasiorowicz and P. Kaus. While they should be given full credit for co-authorship of this work, only I am responsible for the errors herein. I want to thank the Aspen Center for Physics for their hospitality during much of this work.

REFERENCES

1. A. A. Migdal, Central Research Institute preprint KFKI-1977-19, Budapest.
2. A. A. Migdal, Ann. of Physics (N.Y.) 110, 46 (1978).
3. S.-Y. Chu, B. R. Desai, and P. Kaus. Phys. Rev. D 16, 2631 (1977).
4. P. Kaus and A. A. Migdal, Ann. of Physics (N.Y.) 115, 66 (1978).
5. D. Beavis, S. -Y. Chu, and P. Kaus, University of California Riverside preprint UCR-79-3.
6. G. 'tHooft, Nucl. Phys. B72, 461 (1974); G. Veneziano, Nucl. Phys. B74, 365 (1974).
7. S. -Y. Chu, private communication.
8. M. M. Nieto, Phys. Rev. A17, 1273 (1978).
9. M. Dine and J. Sapirstein, Phys. Rev. Lett. 43, 668 (1979).
10. G. 'tHooft, Nucl. Phys. B62, 444 (1973).

11. W. A. Bardeen, A. J. Buras, D. W. Duke, and T. Muta, Phys. Rev. D18, 3998 (1978).

12. J. M. Cornwall and G. Tiktopoulos, Phys. Rev. D13, 3370 (1976).

13. See e.g. R. Dashen, proceedings of this conference.

THE U(1) PROBLEM AND ANOMALOUS WARD IDENTITIES

(Presented by Pran Nath)

Pran Nath *)

CERN -- Geneva, Switzerland

and

R. Arnowitt

Northeastern University, Boston, Massachusetts 02146

ABSTRACT

An effective Lagrangian is developed using the 1/N expansion of QCD which includes the anomaly and θ dependent effects of the fundamental Lagrangian. The effective Lagrangian contains the full solution of the U(1) problem so that there is no light singlet pseudoscalar meson in the chiral symmetry limit. Anomalous Ward identities to all n point order and arbitrary q^2 are shown to hold for this Lagrangian. The mass of the η' in the chiral symmetry limit is related to the properties of the vacuum energy in the absence of quarks as for the fundamental Lagrangian. The effective interaction of the strong CP violation is also obtained and is expressed directly in terms of the phenomenological mesic fields.

1. INTRODUCTION

As we all know the U(1) problem has been with us for many years. It was first pointed out by Glashow[1)] in 1967 that in a QCD theory

*) On sabbatical leave from Northeastern University, Boston, MA.

with three light quarks one has nine rather than eight light pseudoscaler particles. This argument was later shown by Weinberg[2] to lead to an upper bound on the mass of the singlet pseudoscalar meson, the η', so that $m_{\eta'} \leq \sqrt{3}m_{\pi}$. A second aspect of the U(1) problem concerns the vanishing of the $\eta \to 3\pi$ decay in the soft pion limit[3]. Both of the above results are of course in contradiction with experiment[4].

There have been several suggestions since the discovery of the U(1) problem regarding its possible resolution[5-7]. In particular it was pointed out by Kogut and Susskind[6] that the necessary ingredient for the resolution was the existence of singular matrix elements of nonguage invariant operators with singularities at $q^2 = 0$. In 1976 it was observed by 't Hooft[7] in his work on instantons that there exists an anomaly in the axial U(1) channel and this led to the hope for a period of time that instantons would resolve the paradox. However, Crewther[8] then showed that the contributions arising from the instantons alone were not sufficient to satisfy the anomalous Ward identities in a normal way. Recently, a very interesting proposal has been made by Witten[9] to achieve a resolution by analyzing the effect of the U(1) anomaly in the framework of the 1/N expansion[10,11]. In the language of the 1/N expansion, the anomaly produces effects which are an order 1/N smaller than the leading terms. However, this is found sufficient to break the singlet-octet degeneracy and produce a significant nonvanishing value for the singlet mass in the chiral symmetry limit[9,12,13].

A second phenomena which is related to the U(1) problem concerns the appearance in QCD of the vacuum angle θ. The vacuum angle introduces in the fundamental Lagrangian a CP violating interaction proportional to θ so that QCD lagrangian involving gluonic variables only has the form

$$L = -\frac{1}{4} F_{\mu\nu} F^{\mu\nu} + \theta \left(\frac{g^2}{32\pi^2}\right) \tilde{F}_{\mu\nu} F^{\mu\nu} \qquad (1)$$

where $\tilde{F}_{\mu\nu} = \frac{1}{2}\varepsilon_{\mu\nu\alpha\beta} F^{\alpha\beta}$, $\varepsilon_{0123} = +1$ is the dual of $F_{\mu\nu}$: we assume also an SU(N) colour so that one may consider the large N limit[10]. A priori, there appears to be no reason to believe that the θ phenomena and the U(1) problem are interrelated. However, this observation is only superficial and there exists a relation between the two phenomena at a deeper level. In fact, Witten[9] has shown that the mass of the η' in the chiral symmetry limit is directly proportional to the double derivative of the vacuum energy $E(\theta)$ in the absence of quarks at $\theta = 0$, i.e.,

$$m_{\eta'}{}^2 = 4N_\ell F_\pi^{-2} \left(\frac{d^2E}{d\,\theta^2}\right)^{\text{no-quark}}_{\theta=0} , \tag{2}$$

where N_ℓ is the number of light quarks (i.e., $N_\ell = 3$).

Next we discuss the work we have done recently[14)] regarding the U(1) problem. Using the 1/N expansion of QCD we have developed an effective Lagrangian formalism which possesses the following properties. First, it contains a full solution of the U(1) problem so that there is no massless η' particle in the chiral symmetry limit. The solution to the η' problem is achieved through the introduction of a new axial four vector anomaly field $K^\mu(x)$ with mesonic interactions proportional to the topological charge density $\partial_\mu K^\mu(x)$ appearing in the Lagrangian. The effective Lagrangian contains the observed meson fields so that it can be used directly for the computation of processes that involve the anomaly effects. The second property that the effective Lagrangian possesses is that it reproduces all the θ properties of the fundamental Lagrangian. In particular the properties of the vacuum energy as a function of the vacuum angle θ is reproduced and one also obtains an effective interaction for the strong CP violation effects.

Of course, one may ask what is gained by using an effective Lagrangian rather than the fundamental Lagrangian and chiral perturbation theory. While the advantages of the effective Lagrangian approach are well known, it is worth while to recount them here in the context of the solution to the U(1) problem. Perhaps the most obvious advantage of the effective Lagrangian in present context is the guaranteed satisfaction of the general n point anomalous Ward identities for all q^2. The previous analyses thus far in this field have been successful in satisfying the anomalous Ward identities only up to two point order and $q^2 = 0$. Further, the framework of the effective Lagrangian allows one to include an arbitrary amount of SU(3) and chiral symmetry breakdown. In contrast in chiral perturbation theory one must sum over many orders of perturbation to approach the experimental values of physical parameters. In an effective Lagrangian the physical parameters assume their experimental values from the very beginning. Finally, in the effective Lagrangian approach one does not need to use the soft pion approximations. There exists a well-defined set of correction terms given by the effective Lagrangian which are consistently omitted in the soft pion limits and which may sometimes be nonnegligible. The use of the effective Lagrangian includes automatically all such corrections over and above the results of the soft pion approximation.

Before we proceed to formulate the effective Lagrangian including the effects of the axial U(1) anomaly we recall first the formulation of an effective Lagrangian in the absence of the anomaly. The construction of such a Lagrangian in terms of the physically observed mesons, and obeying the principles of current algebra and the PCAC condition, has been known for many years[15]. On the other hand, it has been shown recently by Witten[9] that the 1/N expansion of QCD leads one to an effective Lagrangian involving physically observed mesonic interactions. One may also expect this 1/N expansion of QCD to obey current algebra since the fundamental QCD Lagrangian obeys current algebra. This expectation would be rendered invalid only if the current algebra constraints were to mix different orders in 1/N.

To examine the question of whether current algebra obeys the

1/N expansion, one looks at the structure of the current algebra effective Lagrangian where the interaction part of the Lagrangian can be expanded in terms of a sum of three-point, four-point, etc., interactions so that

$$L_I = \sum_n L_n \; , \; L_n = g^{(n)} \; \phi_1 \ldots \phi_n \quad , \tag{3}$$

where ϕ's represent the full set of observed mesic fields appearing in the effective Lagrangian and $g^{(n)}$ is the strength of the nth order interaction terms. The constraints of current algebra determine $g^{(n)}$ so that $g^{(n)} \simeq F^{2-n}$, where F's represent the strength of the interpolating constants that define the matrix elements of the bilinear quark currents between vacuum and one-meson states, e.g.,

$$< o | \, \bar{q} \, \gamma^\mu \gamma_5 \, \frac{1}{2} \, \lambda_a \, q | b(q) > \; = \; i \, q^\mu \, F_{ab} \quad , \tag{4}$$

where the one-particle states refer to the 0^- nonets of mesons. In the 1/N expansion F's obey the relation $F \sim \sqrt{N}$ so that

$$g^{(n)} \sim N^{1- \frac{n}{2}} \quad . \tag{5}$$

Equation (5) implies that current algebra predicts that the three-point mesic vertices behave as $1/\sqrt{N}$ and so on. In the $N = \infty$ limit all the amplitudes vanish and one has a set of stable and noninteracting meson states. These are precisely the constraints required by the 1/N expansion of QCD obtained by Witten[11]. However, we note that the current algebra analysis determines not only the 1/N order of the n point amplitudes correctly but also their relative strengths. We also note that the current algebra effective Langrangian discussed above does not contain gluonic mesons or glueball states. This is consistent with the fact that the couplings of the glueball states to physical mesons are of higher order in 1/N[16].

The discussion in the remaining part of this paper consists of the following topics. First we discuss the structure of the effective Lagrangian including the anomaly effects and satisfaction of the anomalous Ward identities. The satisfaction of the anomalous Ward identities requires the introduction in the Lagrangian of an anomaly field and gives rise to ghost poles in the matrix elements of nonguage - invariant operators though, of course, the ghosts decouple from all physical gauge invariant amplitudes. Next we shall investigate the θ dependence of the vacuum energy and its relation to the η' mass in the absence of quarks. Finally we shall discuss the form of the strong CP violating interaction that arises in the theory.

2. EFFECTIVE LAGRANGIAN WITH ANOMALY EFFECTS

As pointed out by 't Hooft[7] the divergence of the ninth axial current for the fundamental QCD Lagrangian is modified from the usual soft (chiral symmetry) breaking part by an additional term so that

$$\partial_\mu A^\mu{}_a = D_a + 2N_\ell\ (g^2/32\pi^2)\tilde{F}_{\mu\nu}F^{\mu\nu}\delta_{aq}\ , \tag{6}$$

where D_a is the so-called soft part which breaks chiral symmetry and N_ℓ is the number of light quarks. The effect of the anomaly actually turns out to be an order 1/N smaller than the leading meson terms[9]. Even so the anomaly contribution is seen to be the dominant term in chiral symmetry limit and of sufficient strength to break the singlet-octet degeneracy.

To account for the anomaly effect in the effective Lagrangian, one introduces an additional anomaly field $K^\mu(x)$ in the theory. The anomaly field $K^\mu(x)$ is in effect an axial four-vector ghost field defined so that $\partial_\mu K^\mu(x)$ represents the topological charge density, i.e.,

$$\partial_\mu K^\mu = \text{topological charge density.} \tag{7}$$

The effective Lagrangian which contains the anomaly effects is then given by

$$L = \frac{1}{2C}(\partial_\mu K_\nu)^2 + G(\chi_a)\,\partial_\mu K^\mu - \theta\partial_\mu K^\mu$$

$$-\phi_a{}^\mu \partial_\mu \phi_a + \frac{1}{2}\phi_a^\mu\,\phi_{a\mu} - \frac{1}{2}\phi_a \mu_{ab}\phi_b + L_{C.A.}(\chi_A, \partial_\mu K^\mu), \tag{8}$$

where $\chi_a \equiv (\phi_a, \sigma_a)$ represent the 18-plet of $0^\pm$ scalar fields and χ_A represent the full set of all mesic fields. The quadratic part of the 0^- scalar fields ϕ_a has been exhibited explicitly since these enter in an important way in our discussion while $L_{C.A.}$ represents the remaining part of the current algebra Lagrangian. The mass matrix μ_{ab} in Eq. (8) arises due to the breakdown of chiral symmetry, is diagonal in a,b = 1 .. 7 and is allowed to be nondiagonal in the sectors a,b = 8,9. This latter aspect of non-diagonality of μ_{ab} in the 8,9 sectors is found necessary if one is to accommodate features of a broken SU(3) symmetry in the effective Lagrangian with anomaly. The parameter C appearing in Eq. (8) also plays a crucial role in our discussion and it represents essentially the strength of the topological charge.

In analogy with Eq. (6), the PCAC condition that the Lagrangian of Eq. (8) is required to obey has the form

$$\partial_\mu A^\mu{}_a = F_{ab}\mu_{bc}\phi_c + \delta_{aq}\, 2N_\ell \partial_\mu K^\mu \quad . \tag{9}$$

From Eq. (8) we see that $G(\chi_a)\partial_\mu K^\mu$ represents that part of L which depends on the anomaly where $G(\chi_a)$ is some polynomial in $\chi_a \equiv (\phi_a, \sigma_a)$ with nonderivative couplings. Equation (9) then implies that G obey the condition

$$F_{ab}\frac{\partial G}{\partial \phi_b} = -Z_{1abc}\frac{\partial G}{\partial \chi_b}\chi_c - 2N_\ell \delta_{aq} \quad . \tag{10}$$

Here Z_{1abc} is the constant that enters in the axial current A_a^0 [17]:

$$A_a^o(x) = -F_{ab}\chi_{ob} - Z_{1abc}\chi_{ob}\chi_c \quad , \qquad (11)$$

and is calculated in Ref. 22 . Equation (10) may be integrated to arbitrary n point order. Thus to quadratic order one has[18]

$$G = -2N_\ell F^{-1}_{aq}\phi_a + 2N_\ell\phi_a F^{-1}_{ar} Z_{1rbc} F^{-1}_{bq}\sigma_c + f(\sigma^2) \quad , \qquad (12)$$

where a,b,c,r = 1 .. 9 and $f(\sigma^2)$ is a quadratic σ_c.

The essential new element in the effective Lagrangian of Eq. (8) is that in addition to the usual current algebra parts one has now a new set of terms involving the interaction of the topological charge density directly with the meson fields. One may determine the effect of the anomaly on the meson masses by examining the quadratic part of the Lagrangian and determining the free field propagators. For the propagators corresponding to the pseudoscalar fields one finds $\Delta_{ab} = \delta_{ab}(q^2 + m_q^2)^{-1}$, where

$$M_a^2\,\delta_{ab} = \mu_{ab} + 4N_\ell^2\,C\,F^{-1}_{aq}F^{-1}_{bq} \quad . \qquad (13)$$

From Eq. (13) we find that the (bare) chiral symmetry breaking masses represented by μ_{ab} are clothed by an anomaly part to generate the physical masses represented by m_a.

To see in a more transparent way how the anomaly clothing in the mass relation of Eq. (13) resolves the U(1) problem related to light η' we consider the chiral and SU(3) limit of Eq. (13). In the chiral and SU(3) limit one has[19]

$$\mu_{ab} = 0 \;,\; F_{89} = 0 = F_{98} \;,\; F_{99} = \sqrt{N}_\ell F_\pi \quad . \qquad (14)$$

Thus in the chiral symmetry limit $m_{\eta'}$ is non-vanishing resolving the problem of the light η'. Further since $F_\pi \sim \sqrt{N}$ one has from

Eq. (15) that $m_{\eta'}^2 \sim \sim O(1/N)$ for large N as in Witten's analysis[9]

To gain further insight into the structure of the field K^μ responsible for anomaly clothing, one may decompose the gauge invariant axial current into a symmetry current $\tilde{A}^\mu$ and an anomaly current. Thus one has

$$A^\mu_a = \tilde{A}^\mu_a + \delta_{aq} 2N_\ell K^\mu \quad . \tag{16}$$

One may compute next the vacuum to particle state matrix element of the anomaly current using the effective Lagrangian. Thus one has

$$\langle o|\delta_{aq}\ 2N_\ell\ K^\mu|b(q)\rangle = -\ i\ \frac{q^\mu}{q2}\ 4N_\ell{}^2 C\ \delta_{aq} F^{-1}_{bq}, \tag{17}$$

where b(q) represents a 0^- state. We note that the matrix element in Eq. (17) contains a singularity at $q^2 = 0$. It is precisely this singularity that guarantees a nonvanishing topological charge as was first pointed out by Kogut and Susskind in the so-called "seizing" of the vacuum effect in QCD. Further, to see how the symmetry currents which enter into the physical decay processes are effected by the presence of the anomaly, one can compute the matrix element analogous to Eq. (17) for the symmetry current. Again one finds a singularity at $q^2 = 0$ with exactly the same coefficient as in Eq. (17) but this time with an opposite sign. Thus the gauge invariant current correctly has singularity free matrix elements.

The propagator of the K_μ field is somewhat more complex. From the quadratic part of Eq. (8) one may compute $\Delta_{\mu\nu} = i \langle T(K_\mu K_\nu)\rangle$ and obtain

$$\Delta_{\mu\nu} = -C\ \frac{N_{\mu\nu}}{q^2}\ \left(1- \frac{q_\mu q_\nu}{q^2}\ 4N_\ell{}^2 C\ (F^{-1}_{aq})^2 (q^2 + M_a 2)^{-1}\right) \quad . \tag{18}$$

From Eq. (15) we see that the parameter C is positive so that Eq. (18) contains a monopole ghost[20]. It must be noted, however that the ghost poles couple only to the anomaly currents and cancel out in the physical amplitudes since these involve only the divergence of K^μ. Thus only the residual effect of the ghosts remains.

We shall see in the next section that the physical meaning of the parameter C (which has figured prominently in our discussions thus far) is that it represents quantum fluctuations of the topological charge in the chiral symmetry limit. In addition to producting an anomaly clothing in the mass relations, the parameter C also enters in the solutions of the anomalous Ward identities. At the three-point level one obtains from the 8-9 sectors a set of three equations relating the four interpolating constants F_{88}, F_{89}, F_{98}, F_{99} and the parameter C. One has[21]

$$(m_\eta F_{88})^2 + (m_{\eta'} F_{89})^2 = \frac{4}{3}\,(C_K C_\kappa) - \frac{1}{3}\,C_\pi \quad , \tag{19}$$

$$F_{88}(F_{88}+\sqrt{2}\,F_{98})m_\eta{}^2 + F_{89}(F_{89}+\sqrt{2}F_{99})\,m_{\eta'}{}^2 = C_\pi , \tag{20}$$

$$(F_{88}+\sqrt{2}\,F_{98})^2 m_\eta{}^2 + (F_{89}+\sqrt{2}\,F_{99})^2 m_{\eta'}{}^2 = 3C_\pi + 8N_\ell{}^2 C \quad , \tag{21}$$

where $C_\pi = m^2_\pi F^2_\pi$, $C_K = m^2_K F^2_K$ and[23] $C_\kappa = m^2_\kappa F^2_\kappa$. Equations (19) -(21) allow one to solve for the four interpolating constants and C in terms of two angles ϕ and ϕ'. Thus

$$\frac{\sqrt{2}}{3}\,\frac{C}{\sqrt{E_A E_B}} = \sin\,(\phi - \phi') \quad , \tag{22a}$$

$$F_{88} = (\sqrt{3}\,m_\eta)^{-1}\,[\sqrt{2}E_A \mathrm{Cos}\,\phi + \sqrt{E_B}\,\mathrm{Sin}\,\phi'] \quad , \tag{22b}$$

$$F_{98} = (\sqrt{3}\,m_\eta)^{-1}\,[\,-\sqrt{E_A}\,\mathrm{Cos}\,\phi + \sqrt{2}E_B\,\mathrm{Sin}\,\phi'] \; , \tag{22c}$$

$$F_{89} = (\sqrt{3}\,m_\eta{}')^{-1}[-\sqrt{2}E_A\,\mathrm{Sin}\,\phi + \sqrt{E_B}\,\mathrm{Cos}\,\phi'\,] \; , \tag{22d}$$

$$F_{99} = (\sqrt{3}m_\eta{}')^{-1}[\sqrt{E_A}\,\mathrm{Sin}\,\phi + \sqrt{2}E_B\,\mathrm{Cos}\,\phi'] \quad , \tag{22e}$$

where E_A and E_B are defined by

$$E_A = 2\ (C_K + C_{\kappa}) - C_{\pi} + \frac{1}{3} C \quad , \qquad (23)$$

$$E_B = C_{\pi} + \frac{2}{3}\ C \quad . \qquad (24)$$

The angles ϕ and ϕ' are of course totally unconstrained, and one needs additional input such as the branching ratios of the $\eta \to 2\gamma$ and $\eta' \to 2\gamma$ decays. With this additional experimental input one has then a totally determined system. The resulting equations are somewhat involved and must await a full analysis to determine the numerical values of C and all the F_{ab}. However, one may make rough estimates which yield $C \approx C_{\pi}$ and hence $N_{\ell}^{2}C \sim 0(C_K)$. Thus the contribution of C to the Ward identities is nonnegligible. Similarly one finds the value of the off-diagonal elements of the interpolating matrix, i.e., F_{89} and F_{98} are also nonnegligible. Thus one has, for example, $F_{89}\ m_{\eta'}/\ (F_{99}m_{\eta}) \sim \frac{1}{2}$. In summary one finds that the effects of the anomaly on the solutions of the physical quantities is not necessarily small. One reason for this is that the anomaly effect in the Ward identity Eq. (21) is being magnified by a factor of N_{ℓ}^{2}, which is a rather large factor for $N_{\ell}= 3$.

One may also calculate, using the effective Lagrangian, the matrix element of the topological charge density between vacuum and η and η' states. One has then

$$<0|\partial_{\mu}K^{\mu}|\eta a> \ = \ 2N_{\ell}C\ F^{-1}{}_{a9} \quad a= 8,9 \quad . \qquad (25)$$

The matrix element of Eq. (25) provides in some sense a measure of the "gluonic content" of η and η' Thus, for example, in the SU(3) limit since $F^{-1}_{89} = 0$, the matrix element.

$$<0|\partial_{\mu}K^{\mu}|\eta>_{SU(3)\ Limit} =0 \quad . \qquad (26)$$

The result of Eq. (26) is expected since in the SU(3) symmetry limit

π,K and η form an SU(3) multiplet and η should not have any "gluonic content" since π and K clearly do not. Similarly in the SU(3) limit we have for the η' matrix element

$$\langle 0|\partial_\mu K^\mu|\eta'\rangle_{SU(3)\ \text{Limit}} = 2N_\ell C/F_{99} \quad . \tag{27}$$

The η' matrix element clearly is nonvanishing and large, showing the important role the anomaly plays in the structure of η'[24].

We proceed now to a discussion in greater detail of the anomallous Ward identities at higher point level. Of course, if the PCAC condition with anomaly, i.e., Eq. (9), is satisfied at higher point level, the satisfaction of the anomalous Ward identities is then automatically guaranteed. The satisfaction of Eq. (9) at higher point level brings into the effective Lagrangian a set of additional terms proportional to $\partial_\mu K^\mu$. The general form of the functional differential equation which gives rise to the new couplings in the presence of anomaly can be obtained from the old functional differential equation in the absence of anomaly by the following replacement:[25]

$$\mu_a^{\ 2}\ \phi_a \rightarrow \mu_{ab}\phi_b + 2N_\ell\ F^{-1}_{a9}\ \partial_\mu K^\mu \quad . \tag{28}$$

The functional differential equations that result from the replacement of Eq. (28) can be integrated to an arbitrary order to generate the new couplings. Equation (28) guarantees that the new couplings are of the form $G(\chi_A)\partial_\mu K^\mu$, so that only the gauge invariant combination $\partial_\mu K^\mu$ couples to the mesonic fields. This is precisely the structure that is displayed in Eq. (8). We also note that the strength of the new couplings is determined through the PCAC equation in terms of the nonanomaly couplings. If one assumes a linear representation for the scalar fields and limits oneself to no derivatives for the new couplings, one has that the new couplings then obey Eq. (10). It is these additional couplings that guarantee the satisfaction of the anomalous Ward Identities for the higher point functions.

3. $(d^2E/d\theta^2)\theta = 0$ IN QCD AND IN THE EFFECTIVE LAGRANGIAN

We proceed now to discuss the θ dependence of the vacuum energy. For the case of the fundamental QCD Lagrangian one can show in a straightforward fashion that $(d^2E/d\theta^2)_{\theta=0}$ is given by[26)]

$$\left(\frac{d^2E}{d\theta^2}\right)_{\theta=0} = (g^2/8\pi^2)^2 \langle B^i_a(0)B^i_a(0)\rangle - \tau, \tag{29}$$

where B^i_a are the colour magnetic fields $B^i_a = \frac{1}{2}\varepsilon^{ijk}F^c_{jk}$ and τ represents the quantum fluctuations of the topological charge density as is given by

$$\tau = i \int d^4x \; \langle 0|T(\partial_\mu K^\mu(x)\partial_\nu K^\nu(0))|0\rangle \quad . \tag{30}$$

Here $K^\mu(x)$ is

$$K^\mu(x) = (g^2/32\pi^2)\; \varepsilon^{\mu\alpha\beta\gamma}\; A_\alpha{}^a(F_{\beta\gamma}{}^a - \frac{1}{3} g\, C^{abc} A_\beta{}^b A_\gamma{}^c) \tag{31}$$

and is thus normalized so that

$$\partial_\mu K^\mu = (g^2/32\pi^2)\tilde{F}_{\mu\nu}F^{\mu\nu} \quad . \tag{32}$$

The first term on the right-hand side of Eq. (29) arises due to the dependence of the gluon canonical momentum on θ. This piece, however, cancels as one moves the gradients out of the time ordered product in Eq. (30) and one obtains the relation

$$\left(\frac{d^2E}{d\theta^2}\right)_{\theta=0} = -\, q^\mu q^\nu\; \Delta_{\mu\nu}(q)\Big|_{q^2=0} \quad , \tag{33}$$

where $\Delta_{\mu\nu} = i \langle T(K_\mu K_\nu)\rangle$. By explicit calculation, one finds that

Eq. (33) also holds identically for the case of the effective Lagrangian. Further for the case of the effective Lagrangian we can compute explicitly the various quantities that we have dealt with in arriving at Eq. (33). First we compute τ. Using the definition of Eq. (30) and the effective Lagrangian of Eq. (8) we have that

$$\tau = 4N_\ell^2 C^2 \left(\frac{(F^{-1})_{89}^2}{M_\eta^2} + \frac{(F^{-1})_{99}^2}{M_{\eta'}^2} \right) \quad . \tag{34}$$

In the SU(3) and chiral symmetry limits one can show by using the Ward identities of Eqs (15)-(17) that τ has the following limit:[27]

$$(\tau)^{\text{chiral Limit}} = C \quad . \tag{35}$$

Thus the physical meaning of C is that it represents quantum fluctuations of the topological charge density in the chiral symmetry limit. Further, the effective Lagrangian also gives, by using Eq. (33),

$$\left(\frac{d^2E}{d\theta^2}\right)_{\theta = 0} = C - \tau \quad . \tag{36}$$

By comparison of Eqs (29) and (35) one finds that the QCD analogue of C is the quantity $(g^2/8\pi^2)^2 < B^i_a B^i_a >$. Since in the chiral symmetry limit, τ approaches C, one has from Eq. (36) that $(d^2E/d\theta^2)_{\theta = 0}$ vanishes in the chiral limit exactly as happens in QCD.

The explicit form of the effective Lagrangian of Eqs (34) and (36) also allows us to go to the no-quark limit quite easily. This can be done by simply setting $N_\ell = 0$ in Eqs. (34) and (36). One has then

$$(\tau)^{\text{no-quark}} = 0 \quad , \tag{37}$$

$$\left(\frac{d^2E}{d\theta^2}\right)^{\text{no-quark}}_{\theta=0} = C \quad . \qquad (38)$$

Equation (37) represents the vanishing of the quantum fluctuatuion of the topological charge density in the no-quark limit. Substituting the evaluation of C from (38) in Eq. (15) one obtains that the mass on η' in the symmetry limit is given by

$$m_{\eta'}^2 = (4N_\ell/F_\pi^2)\left(\frac{d^2E}{d\theta^2}\right)^{\text{no-quark}}_{\theta=0} \quad . \qquad (39)$$

As stated in the introduction Eq. (39) was first derived by Witten in the QCD framework. Here it follows from the effective Lagrangian in a rather direct manner. As also stated in the Section I, (39) provides the relationship between two different phenomena, i.e., that the nonvanishing of $m_{\eta'}$ in the chiral symmetry limit needed to solve the U(1) problem requires a nonvanishing $(d^2E/d\theta^2)_{\theta=0}$.

4. THE STRONG CP VIOLATION

The strong CP violation effects arise only with a broken chiral symmetry. This is so because in the chiral symmetry limit one can eliminate θ from the fundamental QCD Lagrangian completely by a U(1) rotation. In the presence of a broken chiral symmetry one may still eliminate the θ dependence completely from the gluon sector but now the θ dependence reappears in the quark sector and one obtains [28)]

$$\delta \mathcal{L}_{cp} = \theta\, m_u m_d m_s (m_u m_d + m_d m_s + m_s m_u)^{-1}\, \bar{q} i\gamma^5 q \quad . \qquad (40)$$

The form of Eq. (40), rather than of Eq. (1) with quarks present, is the one found more useful for carrying out chiral perturbation theory for the computation of strong CP violating effects.

The effective Lagrangian exhibits properties similar to those of the fundamental Lagrangian in this respect also. Thus in analogy to the fundamental Lagrangian one can make an identical U(1) transformation to eliminate the θ term in the topological charge density sector, i.e., the term $-\theta\partial_\mu K^\mu$ in Eq. (8) and introduce it in the meson sector. We proceed now to compute the total ∂L_{CP} that arises when we make a U(1) rotation on the symmetry breaking part of the Lagrangian $L_1(\chi_a, \partial_\mu K^\mu)$, which we write in the form

$$L_1(\chi_a, \partial_\mu K^\mu) = L\,(\chi_a) + G(\chi_a)\partial_\mu K^\mu - \theta\,\partial_\mu\,K^\mu \quad , \tag{41}$$

where $G(\chi_a)$ obeys Eq. (10) and $L(\chi_a)$ is the chiral symmetry breaking part of the current algebra Lagrangian and obeys the PCAC condition[29]

$$F_{ab}\frac{\partial L}{\partial \chi_b} = -\,Z_{labc}\frac{\partial L}{\partial \chi_b}\,\chi_c - F_{ab}\mu_{bc}\,\chi_c \quad . \tag{42}$$

The U(1) rotation which we wish to perform is associated with the generator of the symmetry transformation $\tilde{Q}^5_9$ where

$$\tilde{Q}^5_9 + \int d^3x\, \tilde{A}^0_9\,(x) \quad ; \quad \tilde{A}^\mu{}_a = A^\mu{}_a - 2N_\ell K^\mu \delta_{a9} \tag{43}$$

and $A^\mu{}_\alpha$ are the nonet of gauge invariant axial currents. The desired U(1) transformation is induced by the unitary operator

$$U(\theta) = \exp(i\,\frac{\theta}{2N_\ell}\,\tilde{Q}^5{}_9) \quad . \tag{44}$$

It is easily seen that[30]

$$U(\theta)\,L_1\,U(\theta)^{-1} = L_1\,(\chi_a' - \theta(2N_\ell)^{-1}F_{9a}\,, \partial_\mu K^\mu) \ . \tag{45}$$

The U(1) transformation alone, however, is not enough. One must in addition make chiral SU(3) $\times$ SU(3) rotation to conform to the theorems of Dashen[31]. This requires that one minimize the quantity

$$F(\beta_i) = \langle\, 0|V^{-1}(\beta)\;L, V\,(\beta)|0\rangle \tag{46}$$

with respect to β_i , where

$$V(\beta) = \exp\,(i\beta_i\, Q^5_{\ i}) \qquad , \tag{47}$$

and $Q^5_{\ i}$ are the axial charges of SU(3) × (SU(3). It is then easily seen that this procedure generates a δL_{CP} of the form [32)]

$$\delta L_{CP} = \theta\,(2N_\ell\, \bar{\mu}^{-1}_{\ 99})^{-1} F_{9a} \sqrt{Z}_{9a}\, V_0 \qquad . \tag{48}$$

where $V_0 = (\sqrt{Z}\phi)_9$ is the ninth pseudoscalar density and $(\sqrt{Z})_{ab}$ is the wave-function renormalization matrix of Glashow and Weinberg[33] . The quantity μ_{ab} appearing in Eq. (48) is defined by

$$\bar{\mu}_{ab} = (Z^{-\frac{1}{2}})_{ac}\, \mu_{cd} (Z^{-\frac{1}{2}})_{db} \qquad . \tag{49}$$

From Eq. (48) we see that the effective CP violating interaction that results after elimination of the θ dependence from the topological charge density sector consists of a linear term proportional to the 9th pseudoscalar density. In analogy with the fundamental Lagrangian the CP violating terms vanish in the limit of chiral symmetry. Further, since one now has the form of the CP violating interaction both in the fundamental as well as the effective Lagrangian formalisms one can obtain a relation among parameters appearing in the two formalisms by equating the vacuum matrix element $<0|\,[\tilde{Q}^5_9, \delta L_{CP}]\,|0>$ in the two formalisms. For $N_\ell = 3$ one obtains

$$(\bar{\mu}^{-1}_{\ 99})^{-1} (\sqrt{Z_{9a}}\, F_{9a})^2 = 12 m_u m_d m_s (m_u m_d + m_d m_s + m_u m_s)^{-1} <0|\bar{q}q|0> . \tag{50}$$

Equation (50) represents a sum rule for the chiral mass matrix of the η and η'.

5. CONCLUDING REMARKS

In summary we have obtained from the 1/N expansion of QCD a realistic form of an effective Lagrangian involving the QCD U(1) anomaly and the θ dependent effects. This Lagrangian automatically contains the full solution of the U(1) problem. Further, since the Lagrangian contains physically observed meson states such as π, ρ, A_1 etc., one can use it directly for the computation of physical processes. The problems that remain now concern detailed computations. We mention here two such applications.

The first, and perhaps the most interesting, concerns the application of the method developed above to compute the neutron electric dipole moment without recourse to the bag[28] or the soft pion approximations[34]. Such a calculation is of interest since the weak contribution to the dipole moment in the standard Kobayashi-Maskawa theory[35] is very small, i.e., of the order of 10^{-30} cm. This value is, of course, several orders of magnitude smaller than the present experimental limit of $\leq 10^{-24}$ cm. for the magnitude of the dipole moment. Thus an observation in excess of the weak dipole moment prediction might indicate a real strong interaction θ effect, provided of course the standard Kobayashi-Maskawa mechanism is right. The second application concerns an improved calculation of the theoretical estimate of $\eta \rightarrow 3\pi$ decay. One expects current algebra calculations to be accurate to within 10-15% and so this decay represents a strong test of the validity of the above theoretical ideas.

Finally, after completion of this work we have learned of some work, in the same spirit as discussed here, by other authors[36,37]. The analyses of Refs 36 and 37 also satisfy anomalous Ward identities to arbitrary order. However, these analyses are carried out in the framework of the sigma models and do not discuss the solutions in the general current algebra framework. A second essential difference with the work of Ref. 36 lies in their treatment of the anomaly field, which only enters in the form $\partial_\mu K^\mu$ in the Lagrangian. Consequently, the formalism of Ref. 36 involves variables K^i which are dynamically not determined.

ACKNOWLEDGEMENTS

This work was supported in part by the National Science Foundation.

REFERENCES

1. S.L. Glashow, in "Hadrons and their Interactions", Academic Press, Inc. (New York 1968)p. 83
2. S. Weinberg, Phys. Rev. D11 (1975) 3583.
3. D.G. Sutherland, Phys. Letters 23 (1966) 384.
4. For a review see "Status of the U(1) Problem", by R.J. Crewther, Nuovo Cimento 2 (1979) No. 8 p. 63.
5. H. Fritzsch and M. Gell Mann in Proceedings of the International Conference on Duality and Symmetry in Hadron Physics, edited by E. Gotsman (Weizman Science Press, Jerusalem, 1971)
6. J. Kogut and L. Susskind, Phys. Rev. D11 (1975) 3594.
7. G. 't Hooft, Phys. Rev. Letters 37 (1976) 8; Phys. Rev. D14 (1976) 3432.
8. R.J. Crewther, Phys. Letters 70B (1977) 349.
9. E. Witten, Nucl. Phys. B156 (1979) 269.
10. G. 't Hooft, Nucl. Phys. B72 (1974) 461.
11. E. Witten, Nucl. Phys. B160 (1979) 57.
12. G. Veneziano, Nucl. Phys. B159 (1979) 213.
13. P. Di Vecchia, Phys. Letters 85B (1979) 357.
14. P. Nath and R. Arnowitt, "The U(1) Problem: Current Algebra and the θ vacuum", NUB Nr. 2417 (1979).
15. J. Wess and B. Zumino, Phys. Rev. 163 (1967) 1727;
    R. Arnowitt, M.H. Friedman and P. Nath, Phys. Rev. Letters 19 (1967) 1085;
    J. Schwinger, Phys. Letters 24B (1967) 473;
    C.G. Callan, S. Coleman and B. Zumino, Phys. Rev. 177 (1968) 2239, 2247.
16. While the meson-meson-meson coupling is of order $1/\sqrt{N}$, the meson-meson- glueball coupling is of order 1/N.

17) Terms depending on other field variables in Eq. (11) are suppressed.

18) F^{-1}_{ab} is the matrix inverse of F_{ab} .

19) The last relation in Eq. (14) is obtained in the large N limit, See Ref. 9 .

20) A similar monopole ghost was also observed by Veneziano in his analysis of Ref. 12 .

21) Equations (19)-(21) were first obtained without anomaly effects in Ref. 22 . Modifications due to anomaly were considered in Ref. 8 . Equations similar to Eqs (19)-(21) but involving different interpolating constants were obtained by H. Goldberg, Northeastern University preprint NUB Nr. 2411 (1979).

22) R. Arnowitt, M.H. Friedman, P. Nath and R. Suitor, Phys. Rev. D3 (1971) 594.

23) This relation is obtained under the assumption that the κ field is linearly realized in the strangeness changing vector sector. More generally C_κ measures the breakdown of SU(3) symmetry and can be obtained from the 0^+ strangeness changing spectral function.

24) Even in the SU(3) limit it is inappropriate to construe η' as a glueball since the right-hand side of Eq. (27) vanishes when $N_\ell = 0$, i.e., when the no-quark limit is taken.

25) Thus the general functional differential equation in the presence of anomaly can be obtained by making the replacement of Eq. (28) in Eq. (6.4) of Ref. 22 .

26) This result is established most easily in the temporal guage. See Ref. 9 .

27) The SU(3) and chiral symmetry limit of Eq. (34) is somewhat subtle since one must evaluate F^{-1}_{89}/m_η. Both F^{-1}_{89} and m_η approach zero in these limits and one can verify that the ratio also vanishes.

28) V. Baluni, Phys. Rev. D19 (1979) 2227.

29) For the sake of simplicity we restrict our discussions here to

the case where $L(\chi_a)$ is independent of the gradients of χ_a so that $L = -\frac{1}{2}\chi_a\mu_{ab}\chi_b + \ldots$ Under this restriction L obeys Eq. (42) rather than the more general form which is Eq. (6.4) of Ref. 22 .

30) We note that $\partial_\mu K_\nu$ is invariant under U(1) rotations. More generally it is also invariant under U(3) × U(3) rotations.

31) R. Dashen, Phys. Rev. D3 (1971) 1879.

32) We have used the property that $V^{-1}(\beta)\chi_a V(\beta) = \chi_a + \beta_i \times [Z_{1iab}\chi_b + F_{ia}]$.

33) S. Glashow and S. Weinberg, Phys. Rev. Letters 20 (1968) 224.

34) R.J. Crewther, P. Di Vecchia, G. Veneziano and E. Witten, Phys. Letters 88B (1979) 123.

35) M. Kobayashi and K. Maskawa, Prog. Theor. Phys. 49 (1973) 652.

36) C. Rosenzweig, J. Schechter and G. Trehern, Syracuse Preprint SU-4217-148 (1979); P. Di Vecchia and G. Veneziano, Preprint CERN TH. 2814 (1980). We thank G. Veneziano for informing us of this work.

37) E. Witten, Private communication. We thank E. Witten for informing us of his work.

RENORMALIZING THE STRONG-COUPLING EXPANSION FOR QUANTUM FIELD THEORY: PRESENT STATUS

(Presented by Fred Cooper)

Carl M. Bender
Washington University, St. Louis, Missouri 63130

Fred Cooper
Los Alamos Scientific Laboratory
University of California, Los Alamos, New Mexico 87545

G. S. Guralnik
Brown University, Providence, Rhode Island 02912

Ralph Roskies
University of Pittsburgh, Pittsburgh, Pennsylvania 15260

David Sharp
Los Alamos Scientific Laboratory
University of California, Los Alamos, New Mexico 87545

I. INTRODUCTION

In this talk we review the progress we have made in determining the renormalized strong coupling expansion in quantum field theory. We restrict ourselves here to $\lambda\phi^4$ field theory in d-dimensions. Our starting point is the lattice version of the path integral representation for the Green's functions.

We first determine the unrenormalized strong coupling expansion on the lattice. This is a double series in the dimensionless parameters $x = \frac{M_0^2 a^2}{\sqrt{\lambda a^{4-d}}}$ and $\varepsilon = \frac{1}{\sqrt{\lambda a^{4-d}}}$ where a is the lattice spacing and M_o is the bare mass. We derive a simple set of graphical rules for determining the Green's functions of the theory as a power series in x and ε. For finite field theories (d < 2) we present a scheme for extrapolating to zero lattice spacing ($a \to 0$). This scheme determines a sequence of approximants to the unrenormalized strong coupling series in the form of a series in powers of $M_o^2/\lambda^{2/(4-d)}$.

Next we consider the problem of mass renormalization. For the lattice field theory this consists of eliminating x in favor of $y = \frac{M_R^2 a^2}{\sqrt{\lambda a^{4-d}}}$, where M_R is the renormalized mass. The region of small y is also the critical region, since λ large is equivalent to M_R small. For $d < 2$, $M_R^2 = 0$ occurs when $M_o^2 \to -\infty$, and the path integral is dominated by instantons. Thus, although the unrenormalized strong coupling expansion is valid for $M_o^2 << \lambda^{2/(4-d)}$, the renormalized strong coupling expansion is valid for $M_R^2 << \lambda^{2/(4-d)}$. We obtain the renormalized strong coupling expansion by two methods. One is to extrapolate the renormalized lattice series into the critical regime and then take the continuum limit. The second is to determine exactly the renormalized continuum strong coupling expansion using instantons for $d < 2$. We find our extrapolation procedure for obtaining the continuum limit of the renormalized lattice field theory works well in zero and one dimension when compared with the exact instanton results in the limit $\lambda \to \infty$. We also discuss a second strong coupling expansion for the lattice field theory valid when M_o^2 goes to minus infinity with λ.

II. DERIVATION OF THE LATTICE STRONG COUPLING EXPANSION

The vacuum persistence functional for d-dimensional $\lambda\phi^4$ field theory in the presence of an external source J(x) can be expressed in Euclidean space as a functional integral

$$Z = \langle 0_+ | 0_- \rangle^J = N \int [\mathcal{D}\phi]$$

$$\times \exp\{-\int_{-\infty}^{\infty} d^d x \, [\tfrac{1}{2}(\delta\phi)^2 + \frac{M_o^2}{2}\phi^2 + \frac{\lambda\phi^4}{4} - J\phi]\} . \qquad (2.1)$$

The path integral is defined on a lattice:

$$\vec{x} = \vec{n}a \quad ,$$

$$\int d^d x \to a^d \sum_{\vec{n}} \quad , \quad \phi(x) \to \phi(\vec{n}) \quad ,$$

$$\partial_i \phi(x) \equiv \frac{\phi(\vec{n} + \hat{i}) - \phi(\vec{n})}{a} \quad ,$$

$$\int [\mathscr{D}\phi] = \lim_{\substack{a\to 0 \\ n\to\infty, \vec{n}a=\vec{x}}} \prod_{n} \int d\phi(\vec{n}) \quad . \tag{2.2}$$

Thus, on the lattice

$$Z[J] = \prod_i \int d\phi(i) \; \exp[\, a^{2d} \sum_{n,i} \phi(n)\, G_0^{-1}(n,i) \quad \phi(i) \,]$$

$$\times \exp \left[-a^d \left(\frac{\lambda}{4} \phi(i)^4 + \frac{M_0^2}{2} \phi^2(i) - J(i)\phi(i)\right)\right]$$

where

$$G_0^{-1}(n,m) = \sum_{i=1}^{d} \frac{(\delta_{n',m+\hat{i}} + \delta_{m,n+\hat{i}}) - 2d\delta_{n,m}}{a^{d+2}} \tag{2.3}$$

is the second difference operator. The lattice strong coupling expansion is derived by first treating the kinetic energy as a perturbation and noticing that what remains is a product of ordinary integrals because of the local nature of the interaction.[1-7] That is

$$Z[J] = \exp \left[\sum_{n,m} \frac{\partial}{\partial J(n)} \frac{G_0^{-1}}{2}(n,m) \frac{\partial}{\partial J(m)} \right] \prod_i F[J(i)] \; , \tag{2.4}$$

where

$$F(y) = \int_{-\infty}^{\infty} \exp\left[-a^d\left(\frac{\lambda}{4}x^4 + \frac{M_o^2}{2}x^2 - yx\right)\right] dx$$

$$= \Sigma A_{2n} \frac{y^{2n}}{2n!}$$

and

$$A_{2n} = 2\int_{-\infty}^{\infty} \exp\left[-a^d\left(\frac{\lambda}{4}x^4 + \frac{M_o^2}{2}x^2\right)\right] x^{2n}(a^d)^{2n} dx \ . \quad (2.5)$$

To obtain the lattice strong coupling expansion[5] we expand (2.5) assuming that $M_o^2\, a^2 << \sqrt{\lambda a^{4-d}}$ and obtain[8]

$$A_{2n} = 2(a^d)^{2n}\left(\frac{4}{a^d\lambda}\right)^{\frac{n}{2}+\frac{1}{4}} \Sigma \left(\frac{-M_0^2\, a^2}{\sqrt{\lambda a^{4-d}}}\right)^{\ell} \frac{\Gamma(n/2 + \frac{1}{4} + \ell/2)}{\ell!} \ . \quad (2.6)$$

To determine the vertices of the theory let

$$Z_0[J] = \prod_i \frac{F[J(i)]}{F[o]} = \exp \sum_i \ell n\, (F[J(i)]\,/F[o]) \ ,$$

$$\ell n\, Z_0[J] = \sum_i \ell n\left(\sum_n \frac{A_{2n}}{A_0}\frac{J^{2n}(i)}{2n!}\right) = \sum_i \sum_n \frac{L_{2n}}{2n!} J^{2n}(i) \ . \quad (2.7)$$

The L_{2n} are the vertices of the theory: At each point there is the possibility of exciting 2n bosons with strength $-\frac{1}{(\lambda a^{4-d})^{n/2}}$. They have the following structure

$$L_{2n} = \frac{1}{(\lambda a^{4-d})^{n/2}} \Sigma\, b_k^{(n)} \left(\frac{M_0^2\, a^2}{\sqrt{\lambda a^{4-d}}}\right)^k \ . \quad (2.8)$$

The diagramatic expansion of any Green's function $G(x_1 x_2 \ldots x_n)$ in powers of $\frac{1}{(\lambda a^{4-d})^{n/2}}$ is obtained by applying

$$\exp\left(\Sigma \frac{\partial}{\partial J(n)} G_o^{-1}(n,m) \frac{\partial}{\partial J(m)}\right) \frac{\partial}{\partial J(1)} \cdot \frac{\partial}{\partial J(2)} \cdots \frac{\partial}{\partial J(n)} \tag{2.9}$$

to $Z_0[J]$.

Since $G_o^{-1}(n,m)$ connects nearest neighbors on the lattice, the strong-coupling expansion is also an expansion in the number of interacting lattace sites.

Since the L_{2n} are power series in $\varepsilon = \frac{1}{\sqrt{\lambda a^{4-d}}}$ starting at ε^n, one sees that to each order in powers of ε there are only a finite number of terms contributing to $\ell n Z[J]$. In particular, the ε^n contribution to the 2k-point Green's function $G(x_1 \ldots x_{2k})$ will contain exactly n-k internal lines corresponding to factors of G_0^{-1} connecting vertices L_{2m} with $m < n$.

The first few terms contributing to

$$G_2(ij) = \int \mathcal{D}[\phi]\phi(i)\phi(j)\, e^{-S[\phi]} \tag{2.10}$$

are shown diagramatically in Fig. 1. In Fig. 1 the solid lines represent G_0^{-1}, the dotted lines represent external lines. The series for all the G_{2k} can be organized in terms of diagrams which are one particle irreducible with respect to G_0^{-1}. Call Λ_{2n} the 2n point function which is one particle irreducible with respect to G_0^{-1}. The basic graph-fragments up to 4 G_0^{-1} entering in Λ_{2n} are shown in Fig. 2 along with their evaluation at zero external momenta. We write

$$G_2=\Lambda_2 + \Lambda_2 G_0^{-1} \Lambda_2 + \Lambda_2 G_0^{-1} \Lambda_2 G_0^{-1} \Lambda_2 + \ldots = (-G_0^{-1} + \Lambda_2^{-1})^{-1} \tag{2.11}$$

	Order	Number of G_0^{-1}	Symmetry Number
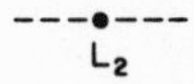	ε	0	1
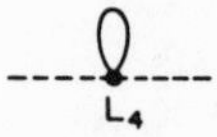	ε^2	1	$\frac{1}{2}$
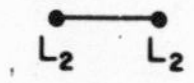	ε^2	1	1
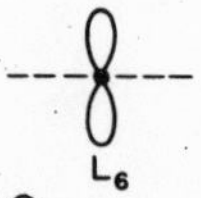	ε^3	2	1/8
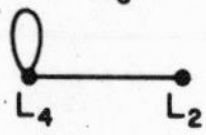	ε^3	2	1/2
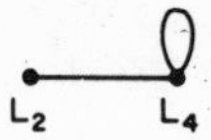	ε^3	2	1/2
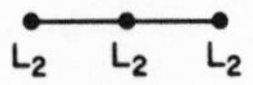	ε^3	2	1
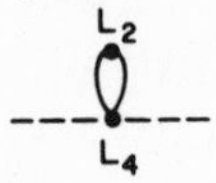	ε^3	2	1/2

Fig. 1: First three orders in the strong coupling expansion of the two point function $G_2(i,j)$.

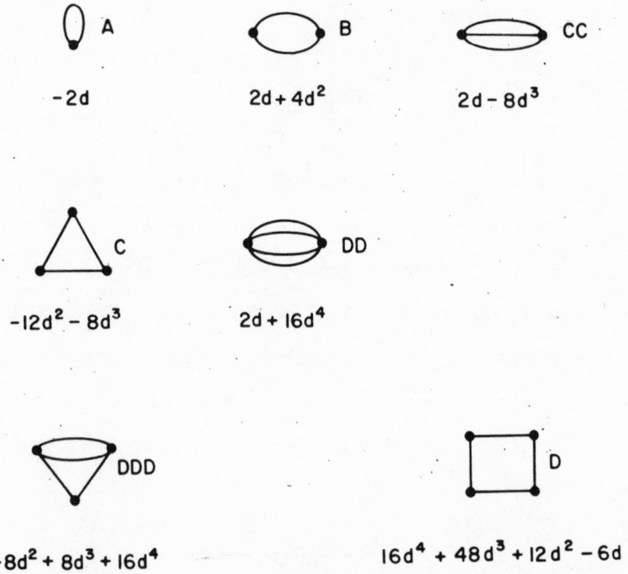
A
-2d
B
2d + 4d²
CC
2d - 8d³
C
-12d² - 8d³
DD
2d + 16d⁴
DDD
-8d² + 8d³ + 16d⁴
D
16d⁴ + 48d³ + 12d² - 6d

Fig. 2: Graph fragments with less than five internal lines G_0^{-1} used in construction $\Lambda_{2n}(p_i=0)$.

Fourier transforming G_2

$$G_2(P) = a^d \Sigma \exp[-i\vec{P} \cdot a(\vec{m} - \vec{\ell})] \quad G_2(\vec{m},\vec{\ell}) ,$$

$$G_2^{-1}(P) = P_E^2 = 1/\Lambda_2(P_E^2) , \tag{2.12}$$

where $\Lambda_2^{(2)} = L_2 + L_4 A + L_4 L_2 B + \frac{L_6 A^2}{2} + L_4 L_4 \, CC + \ldots$

and the A, B, C, etc. are the graph fragments in Fig. 2. Each Λ_{2n} can be written as the sum of products of basic one particle irreducible graph fragments as shown in Fig. 2.

Since the particle spectrum is the zero of the real part of the inverse Green's function in Minkowski space and $-G_0^{-1} = P_E^2 = -P_M^2$, the particle spectrum is the solution of

$$-P_M^2 + \frac{1}{\Lambda_2(-P_M^2)} = 0 . \tag{2.13}$$

For the first zero in G_2^{-1}, which is the renormalized mass we get a series of the form

$$M_R^2 a^2 \varepsilon = 1/(2R) + \sum_{k,\ell}^{k+\ell=n} a_{k\ell}(d)(M_0^2 a^2 \varepsilon)^k (\varepsilon)^\ell , \tag{2.14}$$

where $R = \frac{\Gamma(3/4)}{\Gamma(1/4)} \doteq 0.338$ and $a_{k\ell}(d)$ depend only on the dimension of spacetime and the pure number R.

Similarly, by looking at graphs one can show that

$$G_4 = \Lambda_4(1 + G_0^{-1} G_2)^4 = \Lambda_4(\Lambda_2^{-1})^4 G^4{}_2 . \tag{2.15}$$

Thus, the one-particle-irreducible four-point function Γ_4 satisfies

$\Gamma_4 = G_4(G_2^{-1})^4 = \Lambda_4(\Lambda_2^{-1})^4$, where Λ_4 has the structure

$$\Lambda_4 = L_4 + L_6 A + L_8 A^2 + L_4^2 B + L_6 L_2 B + \ldots \quad (2.16)$$

The renormalized coupling constant λ_R is just $\lambda_R = -Z_3^2 \Gamma^4(P=0)$, where

$$Z_3^{-1} = \left. \frac{dG_2^{-1}(P^2)}{dP^2} \right|_{P^2=0} . \quad (2.17)$$

We find for the dimensionless renormalized coupling constant the following strong-coupling series on the lattice

$$\frac{\lambda_R}{M_R^{4-d}} \varepsilon^{\frac{d}{2}} = \frac{12R^2-1}{16R^4} + \sum_{k,\ell} b_{k\ell}(d) \, (M_0^2 a^2 \varepsilon)^k \varepsilon^\ell . \quad (2.18)$$

Similarly, one can calculate in terms of the graph fragments of Fig. 2 all the coefficients V_{2n} in the expansion of the effective potential about $\phi = 0$:

$$V[\phi] = \sum \frac{V_{2n}}{2n!} \phi^{2n} .$$

The V_{2n}, which are the scattering amplitudes at zero external momentum, are expressed in terms of $\Lambda_{2n}(0)$, which are constructed from the graph-fragments of Fig. 2.

We notice that in the series (2.14) and (2.18) $\varepsilon \to \infty$ as $a \to 0$ and thus every term in these series diverges when the lattice spacing a goes to zero.

III. EXTRAPOLATING TO ZERO LATTICE SPACING

Even in a finite theory such as d=1 - dimensional field theory (quantum mechanics) the lattice strong-coupling series for finite quantities such as M_R or λ_R become infinite term by term as $a \to 0$ (or $\varepsilon \to \infty$). The problem is to extrapolate such series to the continuum after having calculated only a finite number of orders. To examine the series for the renormalized mass, say, we replace a^{-2} by $(\lambda\varepsilon^2)^{2/(4-d)}$ in Eq. 2.14, and obtain

$$\frac{M_R^2}{\lambda^{2/(4-d)}} = f_1(\varepsilon) + f_2(\varepsilon)\left(\frac{M_0^2}{\lambda^{2/(4-d)}}\right) + f_3(\varepsilon)\left(\frac{M_0^2}{\lambda^{2/(4-d)}}\right)^2 + \cdots \quad . \tag{3.1}$$

In $\underline{N}$th order perturbation theory, the $f_n(\varepsilon)$ have the structure

$$f_n^N(\varepsilon) = \varepsilon^{\alpha} \sum_{k=0}^{N} b_k^{(n)} \varepsilon^k \quad . \tag{3.2}$$

For the anharmonic oscillator, a canonical transformation yields

$$H(M_0^2,\lambda) = \lambda^{1/3}\, \bar{H}\, (M_0^2/\lambda^{2/3},1) \quad . \tag{3.3}$$

One expects for d=1

$$\lim_{\varepsilon\to\infty} \lim_{N\to\infty} f_n^N(\varepsilon) = f_n(\infty) \quad , \tag{3.4}$$

where $f_n(\infty)$ is finite. Our problem here is to take $\varepsilon \to \infty$ when we have only a finite number of terms in the series. Eq. 3.4 is a specific example of the problem one always encounters when going to the continuum limit of a lattice theory.

In general, we want to study the series

$$Q(\varepsilon) = \varepsilon^{\alpha} \sum_{n=0}^{\infty} a_n \varepsilon^n \ , \ \alpha \neq 0 \quad , \tag{3.5}$$

which is derived assuming ε is small. We assume $Q(\varepsilon)$ has a finite limit as $\varepsilon \to \infty$. If we truncate the series after the ε^N term, for small ε we can raise the expression

$$Q_N(\varepsilon) = \varepsilon^{\alpha} \sum_{n=0}^{N} a_n \varepsilon^n \qquad (\varepsilon \to 0_+) \tag{3.6}$$

to the N/α power and write the result in the form of a ratio

$$[Q_n(\varepsilon)]^{N/\alpha} \sim \frac{\varepsilon^N}{\sum_{n=0}^{N} b_n \varepsilon^n} \qquad (\varepsilon \to 0_+) \tag{3.7}$$

where the b_n are determined by the a_n.

Now we can take the limit $\varepsilon \to \infty$ and we define

$$Q_N(\infty) = \left(\frac{1}{b_N} \right)^{\alpha/N} , \tag{3.8}$$

as the Nth approximant.

The Q_N in many cases form a convergent sequence of approximants. However, in other cases the sequence resembles the sequence of partial sums of an asymptotic series (it first approaches the exact continuum answer and then veers away). More work needs to be done to understand this sequence[9].

For the anharmonic oscillator the series (3.1) at $\varepsilon = \infty$ is known numerically to be[10]

$$\frac{M_R}{\lambda^{1/3}} = 1.0808 + 0.3399 \frac{M_0^2}{\lambda^{2/3}} + \ldots \quad . \tag{3.9}$$

The first three extrapolants Q_N to 1.0808 are

$$Q_1 = 1.1194, \; Q_2 = 1.1021, \; Q_3 = 1.0973 \; , \qquad (3.10)$$

which appears to converge monotonically.

For the anharmonic oscillator we have calculated twelve terms in the series for the derivative of the ground state energy at $M_0^2 = 0$. The exact result is

$$4\lambda \frac{dE}{d\lambda} \doteqdot 0.569473 \quad \lambda^{1/3} \quad . \qquad (3.11)$$

Better approximants are found evaluating $Q_N(\varepsilon)$ at $\varepsilon = N^{3/2}$ rather than at $\varepsilon = \infty$. The results are:

$$\begin{gathered} 0.477788, \; 0.535128, \; 0.548817, \; 0.554843, \; 0.558673, \; 0.561548, \\ 0.563809, \; 0.565563, \; 0.566845, \; 0.567683, \; 0.568114, \; 0.568195. \end{gathered} \qquad (3.12)$$

We have tested this method of taking the lattice spacing to zero by solving differential equation boundary-layer problems[11]. We first discretize the equation, perturb in the derivative terms and then take the lattice spacing to zero. A boundary-layer problem relevant to $\lambda\phi^4$ field theory is the one-dimensional Euclidean classical field equation known as the kink equation. We obtain this equation by introducing the dimensionless time $\bar{\tau}$, $\mu\tau = \bar{\tau}$ and rescaling the fields by $[\mu^2 = - M_0^2]$

$$\phi(\tau) = \sqrt{\frac{\mu}{\lambda}}\, \tilde{\phi}(\mu\tau) \quad .$$

The classical field equation now becomes:

$$\frac{d^2\tilde{\phi}}{d\bar{\tau}^2} + \tilde{\phi} - \tilde{\phi}^3 = 0 \quad . \tag{3.13}$$

Imposing the boundary conditions $\tilde{\phi}(0) = 0$, $\tilde{\phi}(\infty) = 1$, gives the continuum solution

$$\tilde{\phi} = \tanh^{-1}\left(\frac{\bar{\tau}}{\sqrt{2}}\right) \quad . \tag{3.14}$$

The lattice problem obtained by the ansatz $\tilde{\phi}(\bar{\tau}) \to \phi(n)$ is

$$\varepsilon[\phi(n+1) + \phi(n-1) - 2\phi(n)] + \phi(n) - \phi(n)^3 = 0 \quad , \tag{3.15}$$

where $\varepsilon = 1/a^2$.

We solve this equation by assuming a series for $\phi(n)$ in powers of ε

$$\phi(n) = \Sigma\, \varepsilon^m \phi_m(n) \quad ,$$

with $\phi_0(n) = \begin{cases} 1 & n \geqslant 1 \\ 0 & n = 0 \\ -1 & n \leqslant 1 \end{cases}$. Then on the lattice the derivative of ϕ at the origin is

$$\frac{\phi(1) - \phi(0)}{a} = \varepsilon^{1/2} \sum_{n=0}^{m} b_n \varepsilon^n \quad . \tag{3.16}$$

The exact value of the first derivative of (3.14) at $\bar{\tau} = 0$ is $1/\sqrt{2} \doteq 0.7071$. We have been able to calculate over 40 terms in the series for the derivative at the origin (3.16). In this case the extrapolants whether evaluated at $\varepsilon = \infty$ or $\varepsilon = N$ seem to be a non-converging sequence. The first ten extrapolants are 1, 0.8409, 0.7819, 0.75724, 0.74076, 0.73121, 0.72393, 0.7190, 0.71515, 0.71231. The extrapolants continue to decrease, eventually undershoot reaching a minimum at $Q_{24} = 0.70198$. Then they turn around and increase and at N=41, $Q_{41} = .7080$ and they have again passed the exact answer.

We have tried to understand the divergence structure of the unrenormalized theory in the strong coupling regime for $d \geqslant 2$ by using this same extrapolation procedure. If we consider the derivative of the ground-state energy density for $M_0^2 = 0$, then

$$4\lambda \frac{dE}{d\lambda} = \delta(0) - \int d^d y G_2(x,y) G_0^{-1}(y,x) = \lambda\, f(\varepsilon) \ .$$

For d=2, the sequence of Q_N involves taking fractional roots of negative numbers after N=3, and returns imaginary answers. Thus the extrapolants are clever enough to know that $f(\varepsilon)$) does not have a finite limit as $\varepsilon \to \infty$. In fact they seem to be able to determine the actual manner of divergence (here logarithmic) of $f(\infty)$.

If one divides the series for $4 \frac{dE}{d\lambda}$ by the divergent series for $\ell n(1 + \varepsilon)$ one gets a seemingly convergent sequence of extrapolants when evaluated at $\varepsilon_N = N$.[12] We find for $4(dE/d\lambda/\ell n(1 + \varepsilon)$ that the $Q_N(\varepsilon_N)$ are 0.3121, 0.3639, 0.3766, 0.3858, 0.3932, 0.3987, 0.4031, 0.4069, 0.4102, 0.4132, 0.4161, 0.4189.

Thus, it seems that the extrapolants Q_N provide in many cases quantitatively good approximations to the correct lattice limit. In some cases the Q_N converge very slowly (solution of the diffusion equation) or are asymptotic (solution to the kink equation) and it is clear that more work must be done in understanding these extrapolations. What is comforting is that the Q_N are smart enough to know when one assumes the wrong physics such as an infinite quantity being finite.

IV. OBTAINING THE RENORMALIZED STRONG COUPLING EXPANSION [d < 4]

In finite theories the unrenormalized strong coupling expansion is a series in $M_0^2 a^2 / \sqrt{\lambda a^{4-d}}$ and ε on the lattice and a series in $M_0^2/\lambda^{(2/(4-d))}$ in the continuum. Physical mass renormalization consists of eliminating the unphysical bare mass parameter M_0 in favor of M_R, the position of the lowest pole in the 2-point Minkowski Green's function (the physical energy of the first excited state minus the ground state energy). In this paper we use intermediate mass renormalization defining the renormalized mass as $Z_3 G_2^{-1}(p^2=0)$. For d ⩾ 2, mass renormalization is a necessity since physical quantities such as λ_R are not finite functions of M_0^2 as the lattice spacing a goes to zero. What we are interested in is how physical dimensionless quantities behave for small values of $M_R^2 a^2/\sqrt{\lambda a^{4-d}}$ on the lattice and for small values of $M_R^2/\lambda^{(2/(4-d))}$ in the continuum. It is clear that the regime of large λ, fixed M_R is equivalent to the regime fixed λ, $M_R \to 0$, and it is thus the regime where, on the lattice, the correlation length $\xi \equiv 1/(M_R a)$ is going to infinity. One expects the possibility that certain quantities become singular as $\xi \to \infty$. Thus, obtaining the renormalized lattice strong coupling expansion means we are interested in the lattice field theory near the critical regime ($M_R^2 a^2/\sqrt{\lambda a^{4-d}} << 1$). However, if we look at equation 2.14 we notice that where the unrenormalized strong coupling series is valid, $M_0^2 a^2/\sqrt{\lambda a^{4-d}} << 1$, $M_R^2 a^2/\sqrt{\lambda a^{4-d}}$ is of order 1, and even in the continuum strong coupling series for M_R in the anharmonic oscillator (Eq. 3.9) we notice that $M_R/\lambda^{1/3} \to 1.08$ for $M_0^2/\lambda^{2/3} << 1$. Thus the place where our unrenormalized strong coupling expansion is valid is far from the regime where the renormalized dimensionless parameters $M_R^2 a^2/\sqrt{\lambda a^{4-d}}$ or $M_R^2/\lambda^{2/(4-d)}$ are small. Thus we have two choices. One is to try to perform mass renormalization by inverting Eq. (2.14) to obtain a series of the form ($\varepsilon = 1/\sqrt{\lambda a^{4-d}}$)

$$M_0^2 a^2 = \sum_n b_n(\varepsilon,d)\left(M_R^2 a^2 - \frac{1}{2R\varepsilon} + \dots\right)^n , \qquad (4.1)$$

and try to extrapolate the series in $(M_R^2 a^2 - \frac{1}{2R\varepsilon} + \dots)$ to the regime $M_R^2 \to 0$ taking into account the possibility of singular behavior.

The second method is to try to obtain a different unrenormalized strong-coupling expansion which is valid near the critical region. For example in the anharmonic oscillator we know that the regime where $M_R \to 0$, λ fixed, (which is equivalent to M_R fixed, $\lambda \to \infty$) is the regime where the first excited state and the ground state become degenerate. This clearly occurs in a double well with $M_0^2 \to -\infty$. The almost infinite double well potential has its physics determined by instantons, and for the lattice field theory one is tempted to make the change of variables $M_0^2 = -\mu^2(\lambda/\mu^{4-d})$ so that $M_0^2 \to -\infty$ with λ. In the new variable μ^2 a steepest decent calculation about the point $\phi(n) = \pm\mu^{d/2-1}$ leads to a new lattice strong coupling expansion in the variables

$$\left(\frac{\mu^{4-d}}{\lambda}\right)\frac{1}{(\mu a)^d} \quad , \quad [\mu a]^{2-d} \quad . \tag{4.2}$$

For $d < 2$ this new expansion has the property that $M_R \to 0$ as $\mu \to 0$ so that one is already near the critical regime when λ is large, μ fixed, using the strong coupling expansion valid for the double well.

First let us discuss the method of extrapolation of the mass renormalization formula valid for

$$\frac{M_0^2 a^2}{\sqrt{\lambda a^{4-d}}} \ll 1 \ .$$

First look at the trivial case of field theory in zero dimensions (no kinetic energy) where one does not have to deal with the zero lattice spacing limit. Defining

$$\frac{1}{M_R^2} = L_2(d{=}0) \quad , \quad \frac{\lambda_R}{M_R^4} = -L_4(d{=}0)\, M_R^4 \quad , \tag{4.3}$$

we obtain for the unrenormalized series for M_R^2 and λ_R

$$\frac{M_R^2}{\sqrt{\lambda}} = \frac{1}{2R} + \frac{(4R^2-1)}{8R^2}\frac{M_0^2}{\sqrt{\lambda}} - \frac{(8R^2-1)}{32R^2}\left(\frac{M_0^2}{\sqrt{\lambda}}\right)^2 + \ldots \quad , \tag{4.4}$$

$$R \doteq 0.338 \;, \quad \frac{1}{2R} \doteq 1.4793 \;,$$

$$\frac{\lambda_R}{M_R^4} = 0.811 + 0.278 \frac{M_0^2}{\sqrt{\lambda}} + 0.0349 \left(\frac{M_0^2}{\sqrt{\lambda}}\right)^2 + \ldots \quad . \tag{4.5}$$

We have calculated 15 terms in the convergent power series in $M_0^2/\sqrt{\lambda}$. We notice that $M_R^2/\sqrt{\lambda} \simeq 1.48$ when $M_0^2/\sqrt{\lambda} = 0$. Here mass renormalization consists of inverting equation (4.4) to obtain

$$\frac{M_0^2}{\sqrt{\lambda}} = -1.68 \left(\frac{M_R^2}{\sqrt{\lambda}} - \frac{1}{2R}\right) + 0.332 \left(\frac{M_R^2}{\sqrt{\lambda}} - \frac{1}{2R}\right)^2 - 0.1778 \left(\frac{M_R^2}{\sqrt{\lambda}} - \frac{1}{2R}\right)^3 \ldots \tag{4.6}$$

Inserting Eq. (4.6) into Eq. (4.5) we obtain the renormalized series

$$\frac{\lambda_R}{M_R^4} = 0.811 - 0.469 \left(\frac{M_R^2}{\sqrt{\lambda}} - \frac{1}{2R}\right) + 0.192 \left(\frac{M_R^2}{\sqrt{\lambda}} - \frac{1}{2R}\right)^2 + \ldots \;, \tag{4.7}$$

which is clearly valid for $M_R^2/\sqrt{\lambda} \simeq 1/(2R)$. If we evaluate the series for λ_R/M_R^4 truncated at N terms at $M_R^2/\sqrt{\lambda} = 0$, we get the following series of approximations to λ_R^*, the dimensionless renormalized coupling constant at $\lambda = \infty$: .811, 1.505, 1.924, 2.148, 2.264, 2.323, 2.352, 2.362, 2.360, 2.351, 2.334, 2.311, 2.284, 2.253, 2.217, 2.178. (4.8)

These numbers reach a maximum at 2.362 and then steadily decrease, thereby resembling the approximants for the kink equation. The maximum is within 15% of the exact answer for $\lambda_R^* = 2$ which we will derive by the double well method below.

We can also try to estimate the way in which $M_0^2 \to -\infty$ as $M_R^2 \to 0$ ($\xi \to \infty$). To do this we assume a power law singularity[13]

$$\lim_{M_R \to 0} M_0^2 = -A[M_R^2]^{-\gamma} , \tag{4.9}$$

so that

$$\lim_{M_R \to 0} M_R^2 \frac{d\ell n M_0^2}{dM_R^2} = -\gamma \tag{4.10}$$

(γ is a critical index). Thus, we consider the series ($Z = M_R^2/\sqrt{\lambda} - 1/(2R)$)

$$[Z + \frac{1}{2R}] \frac{d\ell n M_0^2}{dZ} = \frac{1.479}{Z} + 0.708 + 0.0576Z - 0.002Z^2 + \ldots . \tag{4.11}$$

Direct evaluation of the series truncated at M terms and evaluated at $Z = -1/(2R)$ yields the following values for γ:

$$1,\ 0.291,\ 0.337,\ 0.422,\ 0.466,\ 0.517,\ 0.574,\ 0.639,\ 0.714,\ 0.7904,\ 0.8786,\ 1.073 , \tag{4.12}$$

which are all in the vicinity of the exact answer of 1, which we will obtain later by the double well expansion. We have not tried to improve on these results by first "Pade'ing" the series in Z before extrapolating.

Thus, in this simple case we see that we can get qualitatively good results by naively extrapolating our renormalized series valid near $M_R^2/\sqrt{\lambda} \sim 1$ directly to the point $M_R^2/\sqrt{\lambda} = 0$.

For $d > 0$ the series for λ_R/M_R^{4-d} has the form after mass renormalization (see Eq. 4.1)

$$\frac{\lambda_R}{M_R^{4-d}} = \sum_n c_n(\varepsilon,d) \left[M_R^2 a^2 - \frac{1}{2R\varepsilon} + \ldots \right]^n . \tag{4.13}$$

Our strategy for calculating λ_R^* here is to first take $M_R^2 a^2 \to 0$, obtaining a series of the form

$$\frac{\lambda_R}{M_R^{4-d}} = \varepsilon^{\alpha(d)} \Sigma a_n(d)\varepsilon^n \quad , \tag{4.14}$$

where $\varepsilon = 1/\sqrt{\lambda a^{4-d}}$ and α and a_n just depend on the dimensionality of space-time. This is a series having the exact form in (3.5) and we take the lattice spacing to zero by looking at the sequence of extrapolants $Q_N(\infty)$ where N is the order of perturbation theory. For d = 1 and d = 2 this strategy produces an apparently convergent sequence of extrapolants for λ_R^*.

For d = 1 we get the sequence

$$2.424, \quad 3.678, \quad 4.644, \quad 5.251, \quad 5.624, \quad 5.855, \quad \ldots \tag{4.15}$$

which appears to converge to 6.

For d = 2 we obtain the sequence

$$4.743, \quad 8.510, \quad 11.758, \quad 14.173, \quad 15.882, \quad 17.130 \quad . \tag{4.16}$$

By performing a calculation which is very differently organized, G. Baker and J. Kincaid[14] have computed these quantities using 12 terms in a high temperature expansion and have obtained 6 for d=1 14.5 for d=2, which qualitatively agrees with these results.

For d=3, however, we do not obtain a meaningful sequence for λ_R^* by first taking $M_R^2 = 0$ and then taking $a \to 0$. Baker suggests that this is because in d=3, λ_R^* is sensitive to the path one takes in trying to take both M_R and a to zero. In his high temperature expansion approach it was necessary to have $M_R \to 0$ gently with a to get sensible answers (he claims $\lambda_3^* = 25$). (See Ref. 18 for the result of our double well calculation, which avoids this problem).

It thus appears that more work needs to be done in trying to understand how to do double extrapolations ($a \to 0$, $M_R \to 0$) starting from the unrenormalized strong coupling expansion. For $d \leqslant 2$ one can avoid the problem of the second extrapolation ($M_R^2 a^2 \varepsilon \sim 1$ to $M_R^2 a^2 \varepsilon \simeq 0$) by considering a second strong coupling expansion which we will now describe.

For $d < 2$ the place where $M_R^2 a^2 \varepsilon \to 0$ is at $M_0^2 \to -\infty$. Thus, we can try to evaluate Eq. 2.5

$$A_{2n} = 2 \int_0^\infty \exp\left[- a^d \left(\frac{\lambda}{4} x^4 + \frac{M_0^2}{2} x^2\right)\right] x^{2n} (a^d)^{2n} \, dx \tag{4.17}$$

in that vicinity by letting $M_0^2 = - \mu^2 \left(\frac{\lambda}{\mu^{4-d}} \right)$. (This change of variables is suggested by work of Isaacson[15] on the anharmonic oscillator.)

Thus, we can write

$$A_{2n} = \left[\frac{2 a^d}{d-2} \frac{\mu^{3-d}}{\lambda} \left(\frac{\partial}{\partial \mu}\right)\right]^n Z_0[\mu^2] \quad , \tag{4.18}$$

where

$$Z_0[\mu^2] = \int_{-\infty}^{\infty} dx \exp\left[- a^d \lambda \left(\frac{x^4}{4} - \frac{x^2 \mu^{d-2}}{2}\right)\right] \quad . \tag{4.19}$$

Since $\lambda \to \infty$ we can evaluate $Z_0[\mu^2]$ by Laplace's method, the minima being at $x = \pm \mu^{(d-2)/2}$. We find that

$$Z_0[\mu^2] = 4\sqrt{\pi} \exp\left[\frac{(\mu a)^d}{4} \frac{\lambda}{\mu^{4-d}} \right] \frac{\mu}{(\mu a)^{d/2}\sqrt{\lambda}} \times \tag{4.20}$$

$$\left[(1 + \frac{3}{4} \left(\frac{\mu^{4-d}}{\lambda}\right) \frac{1}{(\mu a)^d} + \frac{105}{32} \left(\frac{\mu^{4-d}}{\lambda (\mu a)^d}\right)^2 + \frac{3465}{128} \left(\frac{\mu^{4-d}}{\lambda (\mu a)^d}\right)^3 + \ldots \right] .$$

The L_{2n} have structure

$$L_{2n} = [\mu^{d-2}]^n \, \Sigma \, a_\ell^{(n)} \left(\frac{\mu^{4-d}}{\lambda (\mu a)^d} \right)^n \quad . \tag{4.21}$$

Dimensionless quantities have unrenormalized series of the form

$$A\left(\frac{\mu^{4-d}}{\lambda}, \bar{\varepsilon}\right) = \sum_{n=o}^{N} f_n(\bar{\varepsilon}) \left(\frac{\mu^{4-d}}{\lambda}\right)^n , \tag{4.22}$$

where $\bar{\varepsilon}$ is $(\mu a)^{d-2}$ and $f_n(\bar{\varepsilon})$ has the usual form in (3.2).

In zero dimensions one can now directly calculate the renormalized strong-coupling series. Using (4.3) we obtain

$$M_R^2 = \mu^2 \left[1+ \frac{\mu^4}{\lambda} + 4\left(\frac{\mu^4}{\lambda}\right)^2 + 31\left(\frac{\mu^4}{\lambda}\right)^3 + 364\left(\frac{\mu^4}{\lambda}\right)^4 + 5746\left(\frac{\mu^4}{\lambda}\right)^5 + 113944\left(\frac{\mu^4}{\lambda}\right)^6 + \ldots\right] . \tag{4.23}$$

Since $\mu^2 = \frac{-\lambda}{M_0^2}$, we see that as $M_R^2 \to 0$, for λ fixed

$$\frac{M_0^2}{\sqrt{\lambda}} = -\frac{\sqrt{\lambda}}{M_R^2} . \tag{4.24}$$

Thus, $M_0^2 \to -\infty$ as $-1/M_R^2$ in the critical regime, directly giving -1 as the critical index. For $\frac{\lambda_R}{M_R^4}$ we have

$$\frac{\lambda_R}{M_R^4} = 2 - 2\frac{\mu^4}{\lambda} - 6\left(\frac{\mu^4}{\lambda}\right)^2 - 40\left(\frac{\mu^4}{\lambda}\right)^3 - 434\left(\frac{\mu^4}{\lambda}\right)^4 - 6552\left(\frac{\mu^4}{\lambda}\right)^5 - 126412\left(\frac{\mu^4}{\lambda}\right)^6 + \ldots . \tag{4.25}$$

Since $\mu^4/\lambda \to 0$ when $M_R^4/\lambda \to 0$, this series immediately gives $\lambda_R^* = 2$.

Inverting Eq. 4.23 gives the mass renormalization series

$$\mu^2 = M_R^2 \left[1 - \frac{M_R^4}{\lambda} - \left(\frac{M_R^4}{\lambda}\right)^2 - 11 \left(\frac{M_R^4}{\lambda}\right)^3 - 139 \left(\frac{M_R^4}{\lambda}\right)^4 - 2373 \left(\frac{M_R^4}{\lambda}\right)^5 - 50181 \left(\frac{M_R^4}{\lambda}\right)^6 + \ldots \right] . \quad (4.26)$$

For $\frac{\lambda_R}{M_R^4}$ we have obtained the renormalized strong-coupling series by inserting Eq. 4.26 into 4.25 to obtain

$$\frac{\lambda_R}{M_R^4} = 2 - 2 \frac{M_R^4}{\lambda} - 2 \left(\frac{M_R^4}{\lambda}\right)^2 - 14 \left(\frac{M_R^4}{\lambda}\right)^3 - 166 \left(\frac{M_R^4}{\lambda}\right)^4 - 2714 \left(\frac{M_R^4}{\lambda}\right)^5 - 55866 \left(\frac{M_R^4}{\lambda}\right)^6 + \ldots \quad . \quad (4.27)$$

For d=1, the anharmonic oscillator, we find that $\mu^2 \to 0$ as $M_R^2 \to 0$ so the lattice series in μ^2 is already valid in the critical regime. Thus, to obtain λ^* one need only set $\lambda = \infty$ in Eq. (4.22) and extrapolate to $a \to 0$. For $d = 1$, we obtain in the following sequence of approximants to λ^*_1: 4.8990, 5.2274, 5.3920, 5.4926, 5.5611 ... . The exact answer is 6.

For the anharmonic oscillator we expect that the continuum strong-coupling series for λ_R can be obtained directly via an instanton calculation. The reason for this is as follows. Using the canonical transformation[15]

$$x = \frac{\mu x'}{(g-1)^{1/4}} \quad , \quad p = \frac{(g-1)^{1/4}}{\mu} p' \quad , \quad g = \frac{\lambda}{\mu^3} \quad ,$$

Isaacson has shown that

$$H = \mu(g-1)^{1/2} \left[(p')^2 + \nu \left[(x')^2 - \frac{}{2\nu}\right]^2 + \frac{1}{4\nu} \right] , \quad (4.28)$$

where

$$\nu = \frac{g}{(g-1)^{3/2}} \quad .$$

Thus, as $\lambda \to \infty$ with μ fixed, the problem becomes two separate harmonic wells at $x = \pm \frac{1}{\sqrt{2\nu}}$ with almost degenerate energy levels:

$$E_n \sim \mu(g-1)^{1/2}(2n+1) \quad . \tag{4.29}$$

Also[15], as $\lambda \to \infty$ the renormalized mass goes to zero exponentially fast with g

$$M_R \leqslant \mu(g-1)^{1/2} e^{-Bg^{1/2}} \quad , \tag{4.30}$$

where B is an unknown constant.

However, the regime $g = \frac{\lambda}{\mu^3} \to \infty$ with μ fixed can be reexpressed in terms of the original bare mass parameter $M_0^2 = -\lambda/\mu$. The dimensionless coupling g is also

$$g = -\left(\frac{M_0^3}{\lambda}\right)^2 \quad . \tag{4.31}$$

Hence, large g is equivalent to weak coupling in the double-well situation, and Eq. 4.30 can be written

$$\frac{M_R}{\lambda^{1/3}} \leqslant \frac{|M_0|}{\lambda^{1/3}} e^{-B\frac{|M_0^3|}{\lambda}} \quad , \tag{4.32}$$

which is to be compared with the instanton evaluation of the renormalized mass due to Polyakov[16] and Gildener and Patrascioiu[17]:

$$\frac{M_R}{\lambda^{1/3}} = \frac{|M_0|}{\lambda^{1/3}} \frac{16\sqrt{2}}{\pi} \left(\frac{|M_0^3|}{\lambda}\right)^{1/2} \exp\left[-\frac{2\sqrt{2}}{3}\frac{|M_0^3|}{\lambda}\right] \quad . \tag{4.33}$$

Equation (4.33) is derived assuming that $\lambda/|M_0^3| \ll 1$ (dilute gas of instantons).

We see that the regime where the instanton calculation is valid ($\lambda/|M_0^3| \ll 1$) is the place where $M_R/\lambda^{1/3}$, the expansion parameter of the renormalized strong-coupling series, is small. It is also the place where $g = \lambda/\mu^3$ is large.

It is therefore expected that expanding about the classical minimum

$$\phi(\tau) = \left(\frac{\bar{M}_0^{\,2}}{\lambda}\right)^{1/2} \tanh^{-1}\left(\frac{\bar{M}_0(\tau-a)}{\sqrt{2}}\right), \qquad (4.34)$$

where $M_0^{\,2} = -\bar{M}_0^{\,2}$, one can directly evaluate W_4 in the pseudo-particle approximation and determine the exact strong-coupling expansion for λ_R.

To calculate λ^* it is actually not necessary to do a complete instanton calculation. We have

$$<0|\phi(\tau)\ \phi(0)|0> = \sum_n <0|\phi(\tau)|n>\ <n|\phi(0)|0> = \sum_n |a_n|^2\, e^{-(E_n-E_0)\tau} .$$

In the regime of strong coupling we have

$$E_n - E_0 \sim \begin{cases} \mu\sqrt{g}, & n \neq 1 : \\ \mu\, e^{-A\sqrt{g}}, & n = 1 . \end{cases} \qquad (4.35)$$

Thus, as $g \to \infty$ only the $n = 1$ term contributes. Similarly, if we break up the time orderings in the four-point function and insert states we have terms of the form

$$\sum_{n,m,\ell} <0|\phi|n>\ <n|\phi|m>\ <m|\phi|\ell>\ <\ell|\phi|0> . \qquad (4.36)$$

When $g \to \infty$, the only term which contributes significantly to this sum is that for which $n=\ell=1$ and $m=0$. Thus, the four-point Wightman function factors into a product of two-point functions.

When we put in the time ordering theta functions in τ-space and integrate over τ to obtain the four-point function at zero momentum we obtain the exact result

$$\lambda^* = 6 . \qquad (4.37)$$

TABLE I

DIMENSION	(N + 3)	1	2	3	4	5	6
0	2	.811	1.505	1.924	2.148	2.323	2.362
½	3.32335	1.879	2.624	3.137	3.438	3.613	3.715
1	6	2.424	3.678	4.644	5.251	5.624	5.855
1½	11.63173	3.275	5.379	7.079	8.118	8.878	9.336
2	24	4.743	8.510	11.758	14.173	15.882	17.130

It seems that a possible fit for the exact answer for d<2 is $\lambda^* = \Gamma(d+3)$. In Table I, we compare the first six extrapolants in dimensions 0, 1/2, 1, 3/2, 2 with $\Gamma(d+3)$. We see that the extrapolants are an asymptotic sequence: they eventually overshoot the exact answer by about 15% at the maxima, turn around, and ultimately diverge. This is clear in zero and 1/2 dimension and is apparently happening in the higher dimensions as well.

SUMMARY

We have shown that for d < 2 we can extrapolate the original lattice strong-coupling series such as Eq. 2.18 for λ_R to zero lattice spacing and into the regime where $\frac{M_R^2}{\lambda^{2/(4-d)}}$ is small. The extrapolants predict the exact answer for λ^* (the critical coupling constant), which we determine by an instanton calculation, to within 15%. We hope to find a way to improve the 15% accuracy by finding methods to handle the asymptotic nature of the sequence. It is interesting that for d < 2 the renormalized strong-coupling series is directly calculable from instanton physics. This happens because the critical region ($M_R^2 \to 0$) occurs in the infinite double well regime ($M_0^2 \to -\infty$). For d > 2 one expects for the lattice field theory near $M_R^2 = 0$ the following behavior

$$M_R^2 \sim [M_0^2 - M_0^{*2}]^\alpha \quad , \tag{4.38}$$

with M_0^{*2} finite for fixed lattice spacing a . We are able to evaluate L_{2n} analytically for $M_0^2 \simeq 0$ and $M_0^2 \simeq -\infty$. Our present task is to extrapolate our two strong coupling series on the lattice to the critical region and zero lattice spacing to obtain the continuum strong-coupling series. This calculation is partially completed[18] and suggests that the "double well" strong coupling expansion defined by Eq. (4.20) avoids the double extrapolation problem and will allow a straight-forward determination of the renormalized strong coupling series.

ACKNOWLEDGMENTS

We would like to thank the MIT MATHLAB group for the use of MACSYMA, with special thanks to Jim O'Dell. We would also like to thank George Baker and John Kincaid for sharing their expertise on high temperature expansions of $\lambda\phi^4$ field theory, especially with regards to critical phenomena. One of us, G.S. Guralnik, would like to thank the Brown University Materials Research Lab for partial support under its N.S.F. grant. One of us, R. Roskies, thanks the N.S.F. and the rest of us thank D.O.E. for partial financial support.

REFERENCES

1. S. Hori, Nucl. Phys. 30, 644 (1962).
2. H. Kaiser, Zeuthen Preprint PHE 74-11 (1974). Unpublished.
3. B.F.L. Ward, Nuovo Cim. 45A, 1 (1978).
4. P. Castoldi and C. Schomblond, Nucl. Phys. B, 139, 269 (1978).
5. C.M. Bender, F. Cooper, G. Guralnik and D. Sharp, Phys. Rev. D19, 1865 (1979).
6. R. Benzi, G. Martinelli, G. Parisi, Frascati Preprint 1978.
7. G. Baker, Jr. and J.M. Kincaid, Phys.Rev.Lett. 42, 1431 (1979).
8. The high temperature expansion of Ref. 7 is purely a kinetic energy expansion. The A_{2n} are evaluated numerically for various values of a, λ, and $M_0{}^2$.
9. Certain conditions under which the Q_n will not converge to the correct answer having to do with finite saddle points in the function $Q_n(\varepsilon)$ are discussed by R.J. Rivers, Phys. Rev. D(to be published).
10. F.J. Hioe and E.W. Montroll, J. Math. Phys. 16, 1945 (1975).
11. C.M. Bender, F. Cooper, G. Guralnik, H. Rose, and D. Sharp, Advances in Mathematics (to be published).
12. C.M. Bender, F. Cooper, G. Guralnik, R. Roskies, D. Sharp, Phys. Rev. Lett. 43, 537 (1979).

13. We thank G. Baker for explaining to us this method of determining critical indices and for his sharing of his knowledge of this problem prior to publication.

14. G. Baker and J.M. Kincaid (Private Communication).

15. D. Isaacson, Commun. Math. Phys. 53, 257 (1977).

16. A.M. Polyakov, Nucl. Phys. B120, 429 (1977).

17. E. Gildener, and A. Patrascioiu, Phys. Rev. D16, 423 (1977).

18. The double well evaluation of $Z_0[\mu^2]$ of Eq.(4.20) has been used to calculate λ^* in 1, 2, and 3 dimensions. At sixth order perturbation theory we obtain $\lambda_1^* = 5.56$, $\lambda_2^* = 10.62$, $\lambda_3^* = 28.42$. These results are consistent with the results obtained by George Baker using a high temperature series.

ON THE n-p MASS DIFFERENCE IN QCD

Geoffrey B. West

Los Alamos Scientific Laboratory

Los Alamos, New Mexico 87545

I. INTRODUCTION AND REVIEW

With the advent of unified gauge theories, quantum chromodynamics, dynamical symmetry breaking and the like, the old and, as yet, unsolved problem of the n - p mass difference has been cast in a new light. Traditional attmepts to explain the infamous sign reversal have invariably been based upon the assumption that the mass splitting was electromagnetic in origin.[1] However, all attempts along these lines agree with naive Coulomb energy considerations and, therefore, fail. Nowadays one has an equally important origin of isospin symmetry breaking, namely, the quark masses themselves.[2] Thus, the problem of the n - p mass difference becomes intimately related to the problem of quark mass splitting and these presumably arise from the dynamical breaking of an assumed unified symmetry. From this point of view, the smallness of isospin splitting appears accidental, being merely a reflection of the smallness of the current quark masses relative to more typical hadronic mass scales. This view leads to a remarkable conclusion, namely, that the formidable macroscopic consequences of the neutron being heavier than the proton have their origins at energies relevant to the breaking of a grand unified group.

Regardless of the role played by the masses themselves, the conventional electromagnetic contribution has to be reckoned with; indeed, all quark mass splittings have two such contributions, namely those from dynamical symmetry breaking and those from gauge boson radiative corrections. It is not obvious, a priori, which contribution should dominate. Indeed, it would appear that the "sign reversal" of the n - p mass difference (ΔM) is a fortuitous dynamical accident. The work of this paper focuses upon the radiative contribution to ΔM; although we shall talk in terms of the n - p system, the calculation can be viewed as a prototype of any such gauge boson contribution to a self-mass or mass difference.

For the radiative contribution, there are, in general, two issues to be addressed, namely, its finiteness and sign. Before going into details, it is useful to review briefly the physics involved together with some of our conclusions. It is well-known that the purely electromagnetic contribution not only gives the incorrect sign for ΔM but also suffers from being logarithmically divergent if the elementary fields are taken to be point-like. This necessarily means that ΔM is sensitive to short-distance effects, and this has two immediate consequences. The first, emphasized by Weinberg,[3] is that in a unified electroweak theory where the strength of the weak interaction eventually becomes comparable to that of the electromagnetic one can expect the weak interactions to play an important role in considerations of finiteness. Indeed, Weinberg described a class of electroweak models where the built-in symmetry necessarily requires ΔM to be finite. Loosely speaking, the presence of a divergence would normally require the introduction of counter terms whose presence would destroy the symmetry. These models ignore the strong interactions entirely, treating the "hadron" as point-like; however, the second consequence of sensitivity to short distances is that the strong interaction gluon radiative corrections must be taken into account. We know that in QCD this generally leads to a very mild logarithmic softening of a typical quark vertex and, therefore, suggests the possibility that ΔM could be finite independent of the Weinberg mechanism. The degree of convergence (or divergence) will, in this case, be governed by the anomalous dimension of the leading operator occurring in the operator product expansion of the associated virtual Compton amplitude.[4] Since ΔM is a scalar, only scalar operators contribute; the leading such operator is the trace of the energy-momentum tensor, θ, which has <u>no</u> anomalous dimensions and so the naively anticipated dampening is lost and ΔM will, in general, remain divergent. Since the appearance of a preliminary version[5] of this work, a paper by Collins[6] has been published which makes a similar observation. He confines his discussion to the purely e.m. contribution; however, the fact that this divergence is insensitive to gluon corrections (i.e., that it is independent of the strong interactions) means that the potential conspiracy leading to convergence within a purely electroweak theory can still be realized, even in the presence of the strong interactions. Now, suppose that such a mechanism is, indeed, operating so that the leading logarithmic divergence is cancelled, then we still have to ensure that the next-to-leading contribution also converges. This, of course, arises from operators that do have anomalous dimensions and are thus truly sensitive to the strong interactions. A straightforward calculation[5] shows that these contributions are finite provided the number of flavors, n_f, exceeds 67/6 [within the standard $SU(3)_c$ scheme]; this effectively requires $n_f \geqslant 12$. If such a scenario is, indeed, correct, then convergence is exceedingly slow, and this allows the weak interactions to contribute significantly in magnitude to ΔM in addition to simply canceling the leading divergence. Whether this is the origin of a sign reversal is not clear; however, in this picture, the fact that such a possibility exists arises from a crucial interplay between the unified aspects of the electroweak sector and the asymptotic freedom of the strong. Of course, if such a calculation were to leave ΔM divergent, then one would be forced to consider the problem within the broader content of a grand unified scheme in order to take account of all interactions on an equal footing.

Having expressed all this in words, let us indicate the final development. Our starting point is Cottingham's formula[7] for the lowest order e.m. contribution to ΔM:

$$\Delta M = -\frac{\alpha}{8\pi^2}\int \frac{d^4q}{q^2} T_{\mu\nu}(p,q) g^{\mu\nu} \quad . \tag{1}$$

Here, q is the virtual photon momentum, p that of the nucleon, and $T_{\mu\nu}$ the associated virtual Compton amplitude:

$$T_{\mu\nu} = \int d^4x e^{iq.x} \langle p|T[j_\mu(x)\ j_\nu(o)]|p\rangle \quad , \tag{2}$$

with j_μ the e.m. current, and $\nu \equiv p.q/M$. This formula can be naively "derived" by noting that the lowest order contribution to the e.m. energy density is $eA_\mu j^\mu$ with $\Box^2 A_\mu = j_\mu$; equation (1) is simply a momentum space representation of this fact with the appropriate Feynman boundary conditions incorporated.

At this stage, we will not specify whether equation (1) refers to selfmasses or mass differences. As usual,[4] $T_{\mu\nu}$ can be decomposed into its two scalar pieces $T_{1,2}(q^2,\nu)$ whose imaginary parts are the standard structure functions $W_{1,2}(q^2,\nu)$. However, for most of our discussion, it is, in fact, easier to stay with the scalar amplitude $T(q^2,\nu) \equiv T_{\mu\nu}(p.q)\ g^{\mu\nu}$. We shall work with the Wick rotated version of (1), although this is not strictly necessary. It is convenient to introduce new variables $Q^2 \equiv -q^2$ and $y \equiv \nu/Q$ and consider T as a function of Q^2 and the scaling variable $x \equiv Q^2/2\nu$ in terms of which (1) can be expressed as

$$\Delta M = \frac{1}{2\pi}\int_o^\infty dQ^2 \int_{-1}^{1} dy(1-y^2)^{\frac{1}{2}}\ T(Q^2,Q/2y) \quad , \tag{3}$$

where $y \equiv \nu/Q$. There are several well-known immediate consequences of this formula which are worth recording:

a) For a point-like particle $T(Q^2,Q/2y) \to 1/Q^2$, resulting in the well-known logarithmic divergence associated with mass renormalization.

b) The fixed Q^2, large ν behavior of the W_i (Q^2,ν) is controlled by the Regge exchanges in the t-channel:[8]

$$W_1(Q^2,\nu) \to \beta_1(Q^2)\nu^{\alpha_I} \quad ,$$

$$W_2(Q^2,\nu) \to \beta_2(Q^2)\nu^{\alpha_I - 2} \quad ,$$

where the β_i are the residue functions, and α_I is the slope of the trajectory, I denoting the exchanged isospin. Phenomenologically $\alpha_0 \simeq 1$ (corresponding to Pomeron exchange), $\alpha_1 \simeq 1/2$ (the ρ trajectory) and $\alpha_2 < 0$ (reflecting the presumed absence of exotics). Thus, the singlet and isovector amplitudes require at least one subtraction in their dispersion representations, whereas, the isotensor requires none. This observation led Harari[8] to conjecture that this is the reason that the naive "low energy" calculations work for the $\pi^+-\pi^0$ mass difference (I = 2), whereas, they fail for the n - p (I = 1), since these involve an "unknown" subtraction term. We shall return to this point below when discussing the role of QCD. Meanwhile, it should be pointed out that it is not yet clear that Regge phenomenology is necessarily a consequence of QCD.

c) The analyticity of the T_i was exploited by Cottingham in an attempt to express ΔM in terms of the directly measurable W_i. As already mentioned above, the presence of the subtraction term for the I = 1 case destroys this possibility; this can be made explicit by expressing the fixed Q^2 dispersion relations in terms of the scaling variable x:

$$T(Q^2,x) = T(Q^2) - \frac{1}{2}\int_0^1 \frac{dx'^2}{x'^2-x^2} F(x',Q^2) \quad , \tag{4}$$

where $F \equiv 2xW_1+\nu W_2 \equiv 2xF_1+F_2$ and $T(Q^2) \equiv T(Q^2,\infty)$ is the subtraction term. Note, incidentally, that, although $|x|<1$ for the W_i, it remains unbounded for the T_i. It is interesting to examine the convergence of ΔM given the representation (4); a straightforward calculation leads to[9]

$$(\Delta M)_{\text{Div.}} \sim \int^\infty \frac{dQ^2}{Q^2}\left[Q^2 T(Q^2) + \int_0^1 F(x,Q^2)\, dx\right] \quad . \tag{5}$$

The original scaling hypothesis of Bjorken[10] that, for large Q^2, $F(x,Q^2)$ becomes independent of Q^2, clearly leads to a logarithmically divergent contribution. This is, of course, to be expected, since exact scaling is a reflection of point-like behavior. Of course, one still has the subtraction term to deal with; a naive application of the Bjorken-Johnson-Low limit[9,11] shows that this also will lead to a logarithmic divergence. It is amusing to examine the <u>un</u>subtracted (I = 2) case in this light; the analog to (5) reads

$$(\Delta M)_{\text{Div}} \sim \int^\infty \frac{dQ^2}{Q^2}\int_0^1 dx F(x,Q^2) \tag{6}$$

showing that, contrary to one's naive expectation, an unsubtracted dispersion relation does not necessarily lead to a convergent result for ΔM. Therefore, without a further assumption such as $F_{\pi^+}(x) = F_{\pi^0}(x)$, for example, one cannot justify the success of the I = 2 calculations. However, a condition such as this can be expected to be true in the quark model, again because of the absence of exotics, precisely the physics which leads to an unsubtracted dispersion relation ($\alpha_2 < 0$).

II. QCD AND THE DIVERGENCE OF ΔM

The expression, equation (3), shows that the convergence of

ΔM is sensitive to the large Q^2 and large x behavior of $T(Q^2,x)$; whether this is controlled by the light cone and standard operator product expansion is a delicate point. This problem can be circumvented by recalling that the scaling region (i.e., $Q^2 \to \infty$ for x fixed) is sensitive to the light cone so one can use analyticity, as shown explicitly in equation (5), to relate the $Q^2 \to \infty$, fixed y behavior (relevant to ΔM) to the light cone. Unfortunately, however, it is not obvious that a confining theory, such as QCD, will lead to analytic amplitudes. Of course, if this were the case, the predictive power of QCD is considerably reduced since the well-known results for the structure function moments rely crucially on analyticity.[4] It is interesting, therefore, to proceed without involving analyticity explicitly but assume only the validity of the operator product expansion for the relevant region.[4] In momentum space, such an expansion reads

$$T(Q^2,x) \to \sum_{n,a} \tilde{f}_n^a(Q^2) q_{\mu_1} \ldots . q_{\mu_n} \langle p|O_a^{\mu_1 \ldots \mu_n}|p\rangle \quad , \qquad (7)$$

where the $\tilde{f}_n^a(Q^2)$ are c-number coefficients and the $O_a^{\mu \ldots \mu_n}$ are a complete set of operators for the theory; a labels the various quantum numbers. The Q^2 independent matrix elements can be decomposed as follows:

$$\langle p|O_a^{\mu_1 \ldots \mu_2}|p\rangle = A_n^a\, p_{\mu_1} \ldots p_{\mu_n} + p^2 B_n^a\, g_{\mu_1\mu_2}\, p_{\mu_3} \ldots p_{\mu n} + \ldots \ldots \qquad (8)$$

For the lowest twist operators (which dominate the scaling region) such as $\bar{\psi}\lambda_a \gamma^{\mu_1} D^{\mu_2} \ldots . D^{\mu_n}\psi$ the corresponding $\tilde{f}_n^a(Q^2)$ have regular dimension $(\text{mass})^{-2n}$; we therefore, introduce dimensionless coefficients $f_n^a(Q^2) \equiv (Q^2)^n \tilde{f}_n(Q^2)$ since these will satisfy the standard renormalization group equation and, therefore, fall-like powers of $\ell n Q^2$. It is not difficult to see that the nth term in (4) has the following structure:

$$\sum_a Q^{-n} f_n^a(Q^2) \; [A_n^a y^n + p^2 B_n^a \, y^{n-2} + \ldots\ldots\ldots] \quad , \tag{9}$$

the terms in square trackets being a polynomial in y of degree n. This shows that for a given n, <u>all</u> twists contribute equally to ΔM unlike the case of the x- moments of the W_i where only the higher twists are relevant. This is easily confirmed from equation (9) by noting that, since $y = Q/2x$, the term involving A_n^a dominates when $Q \to \infty$ with x fixed. For the divergent part of ΔM, the fact that all twists contribute is, superficially, not a serious one, since only the leading two terms in the series are relevant to the convergence properties of (3). Indeed, equation (3) can be expressed as

$$\Delta\, M = \frac{1}{4} \int^{\infty} dQ^2 \sum_a [A_o f_o^a(Q^2) + \frac{p^2}{4Q^2} f_2^a(Q^2)(A_2^a + 4B_2^a) + 0(1/Q^4)] \tag{10}$$

$$= \frac{1}{4} \int^{\infty} dQ^2 \sum_a [f_o^a(Q^2) \; \langle p|0_a|p\rangle + \frac{f_2^a(Q^2)}{Q^2} \langle p|0_{\mu\mu}^a|p\rangle + 0(\frac{1}{Q^4})] \; . \tag{11}$$

In general, there is, of course, the more serious problem as to whether the OPE applies to the fixed y limit relevant to ΔM; this is the same question as to whether the OPE applies to the boundary regions in the x-plane (here $x \to \infty$). The conventional way around this potential difficulty is to employ the standard fixed Q^2 analyticity properties of the $T_i(Q^2,\nu)$. This will, indeed, justify an expression like (11); however, we wish to emphasize that analyticity in general plays a crucial role.

Now, in equation (11), it is easy to convince oneself that, for a dimensionless $f_o^a(Q^2)$, there are no operators 0_a. There are, of course, higher twist contributions and, in a massive theory, the leading one is (generically) of the form $M\bar{\psi}\psi$. However, this requires $f_o^a(Q^2) \to 1/Q^2$ and it is somewhat neater to consider this as a lowest twist contribution to $f_2^a(Q^2)$ to be associated with the operator $g_{\mu\nu}M\bar{\psi}\psi$. Thus, without loss of generality, we can drop the leading term in (11) and write

$$\Delta M_\infty = \langle p|0^a_{\mu\mu}|p\rangle \frac{1}{4} [\int^\infty dt \sum_a f_2^a(t)] \quad . \tag{12}$$

The subscript ∞ on ΔM indicates that we are keeping only those terms that can lead to a divergence, dropping all explicitly convergent ones. Furthermore, we have expressed the integrand in terms of the dimensionless variable $t \equiv \ln(Q^2/\Lambda^2)$, where Λ is the standard QCD scale parameter.

Now $f_2^a(t)$ satisfies the renormalization group equation and, as such, behaves like $t^{-c_2^a/2b}$ for large t where c_2^a is defined as the coefficient of g^2 in the perturbation expansion of the anomalous dimension of $0^a_{\mu\nu}$ (g being the QCD coupling constant), and -b the coefficient of g^3 in the corresponding expansion of the standard β-function. It is well known that c_2^a does not depend explicitly on the number of quark flavors whereas b does; indeed, $b = (11-2/3n_f)/16\pi^2$ and asymptotic freedom fails for $n_f > 16$. Now, from equation (12), it is clear that for ΔM to be convergent, we require $c_2^a/2b > 1$ and, in principle, this can lead to a lower bound for n_f. Any conclusion, however, depends upon c_2^a and we now turn to a discussion of the operators $0^a_{\mu\nu}$. Typical of the lowest twist operators, besides the aforementioned $g_{\mu\nu}M\bar{\psi}\psi$, are $F_{\mu\alpha}F^{\alpha\nu}$ and $\bar{\psi}\gamma_\mu D_\nu\psi$; for convenience the possible dependence on a is momentarily suppressed. We are, incidentally, using the standard notation in which $F_{\mu\nu}$ is the non-Abelian field tensor and D_μ the corresponding generalized covariant derivative. There is, of course, a combination of these operators which is the stress-energy tensor, $\Theta_{\mu\nu}$, which has no anomalous dimension, since it is the generator of an exact symmetry. Its trace, which is what contributes to (12), is of the form

$$\Theta = \sum_f M_f\bar{\psi}_f\psi_f + F_{\mu\nu}F^{\nu\mu} \quad , \tag{13}$$

where the color indices have been suppressed. Although this actually has to be modified by anomalies,[6] it is only its general structure which is of importance here. In particular, the fact that the gluon term is flavor independent, and, in particular, independent of ψ_f means that there is a piece of ΔM_∞ which is itself a flavor singlet and which has no anomalous dimension (i.e., $C_2 = 0$). Thus, the e.m. contribution to ΔM will remain divergent even in the presence of the strong interactions. Note, incidentally, the generality of this result: it relies solely upon the nonvanishing of the trace of the energy-momentum tensor and so remains valid in <u>any</u> theory!

One might hope that the divergence would cancel for a mass difference; unfortunately, however, this is not so. The gluon contribution to the divergence is of a purely singlet nature and therefore does indeed cancel. The flavor dependent quark contributions, on the other hand, are, through the coefficients f_2, weighted with the squares of their charges, so, even if their masses were degenerate, they do not, in general, cancel. Indeed, each quark flavor (f) contributes a term $\langle p|Q_f^2 M_f^2 \bar{\psi}_f \psi_f|p\rangle$. This is, of course, <u>identical</u> to the lowest order contribution to the electron's self-mass in QED. This fact will have important consequences when we come to consider a unified electroweak theory below. Before doing so, however, there are a few remarks worth emphasizing: a) The singlet contribution arising from the gluonic piece is divergent for the same reason[4] that the area under the scaling curve for $F_2(Q^2,x)$ is independent of Q^2. Indeed, this is clearly reflected in equation (5) where the division between the contributions from the absorptive and real parts roughly corresponds to the division between the quark and gluon contributions. b) Returning to the I = 2 mass differences, we note that the leading operator requires four quark field operators; the added dimension associated with such an operator (over the bilinear combination relevant to I = 0 or 1) leads to an extra power of $1/Q^2$ in $f_2(Q^2)$ and thus ensures convergence. This argument, which is, of course, equivalent to the absence of exotics, supplements

Harari's original observation[8] and ensures that a "naive" calculation is sufficient for I = 2. c) It is, of course, always possible to swallow up the divergences in a redefinition of the renormalized quark masses. This is the procedure that is implicitly carried out in nuclear physics; for example, consider the mass difference between He^3 and H^3. This has two major contributions; one from the n - p mass difference itself and one from the Coulomb interaction energy between nucleons; symbolically:

$$M(He^3) - M(H^3) = (M_p - M_n) + \Delta(\text{Coulomb}) + \ldots. \qquad (14)$$

The use of the physical nucleon masses in this equation is equivalent to using renormalized masses; in other words, the fermion <u>self</u> energies arising from e.m. are assumed to be the origin of the physical n - p mass difference to be used in equation (14). The so-called Coulomb energy contribution is the e.m. energy of the <u>inter</u>nucleon interaction. In the formal language that we have been using, these terms arise from operators that are <u>quadratic</u> in the fermion fields (much as the I = 2 operators are) and, thus, give a convergent calculable contribution. In the parton model, such higher twist terms physically correspond to a coherent contribution to the W_i ($\alpha \sum_{i\neq j} Q_i Q_j$), in contrast to the bilinear operators which correspond to the conventional incoherent parton contribution ($\alpha \sum_i Q_i^2$). It is well-known that for the proton $\Sigma Q_i^2 = 1$, whereas, for the neutron, it is 2/3 (this being the "origin" of the naive conclusion that $M_p > M_n$). It is less well-known[5] that the coherent sum $\sum_{i\neq j} Q_i Q_j$ vanishes for the proton, whereas it is -2/3 for the neutron, again, unfortunately, leading to $M_p > M_n$. Note, however, that this latter result depends on the details of the charge assignments for the quarks.

III. THE UNIFIED ELECTROWEAK CONTRIBUTION AND THE POSSIBLE CONVERGENCE OF ΔM

Thus far, we have shown that the absence of an anomalous dimension for Θ will generally lead to a logarithmically divergent ΔM; in fact, for the nonsinglet contribution, the divergent part of ΔM can be calculated as if the system were composed of point-like partons. It is of interest to consider the next term in the OPE; i.e., the leading nonsinglet operator beyond the quark mass terms in Θ already considered. For example, suppose, for simplicity, that matrix elements of Θ vanished in the I = 1 channel. Then, the convergence of ΔM would be governed by nonsinglet operators of the generic form $\bar{\psi}\lambda_a\psi$. Such operators do, indeed, have anomalous dimensions and are, therefore, sensitive to the gluon content of the theory. Indeed, a standard calculation leads to $c_2 = 4/9\pi^2$; since convergence requires $c_2/2b > 1$, we deduce that even the nonleading term is not guaranteed to converge unless $n_f > 67/6$. Since n_f must be integral, this requires $n_f \geqslant 12$.

Suppose, now, that this condition is satisfied, then, according to equation (12), the deep inelastic contribution to ΔM is of the form

$$(\Delta M)_\infty \sim \alpha \int_{t_o}^{\infty} \frac{dt}{t^p} \quad , \tag{15}$$

where

$$p \equiv \frac{c_2}{2b} = \frac{32}{3(33-2n_f)} > 1$$

and $t_o\ \ell n\ Q_o^2/\Lambda^2$; Q_o represents the onset of the scaling region (~ 1GeV/c) and Λ the QCD scale parameter (~ 300 Mev). Clearly,

$$(\Delta M)_\infty \sim \frac{\alpha}{t_o^{\,p-1}} = \frac{\alpha}{[\ell n\ Q_o/\Lambda]^{p-1}} \quad . \tag{16}$$

This shows explicitly that this contribution is very sensitive to large values of Q^2, so large in fact that, in a unified electroweak theory, the weak contribution could become significant. For n_f near its lower limit, this is even truer; for example, for $n_f = 12$, $p-1 = 5/27$. This notion can be extended to the idea that if the e.m. contribution to ΔM is divergent, then, in a unified theory, all of the interactions are, in principle, of equal importance. Such a viewpoint has been stressed by Weinberg[3] who introduced the concept of a natural symmetry which can be broken in conventional electroweak theories, but in such a way that non-singlet quantities have finite renormalizations. The idea has been illustrated in models where the strong interactions are ignored; basically, the spontaneous symmetry breaking is constrained so that the standard logarithmic divergence in the e.m. self-energy is precisely balanced by a similar divergence in the weak sector. The question naturally arises as to what happens to this mechanism in the presence of the strong interactions. This has, in fact, already been answered in the previous section, for there, we saw that, for mass differences, the logarithmic divergence arises solely from the quark content of Θ and is totally insensitive to the presence of gluons. Indeed, we remarked that the divergent contribution can be calculated by ignoring the strong interactions entirely treating the quarks as structureless partons, as in the Weinberg models. We, thus, conclude that the cancellation of the divergences inherent in a model with natural symmetry remains valid even in the presence of gluons.

If this mechanism is, indeed, operating, then the leading terms, which are sensitive to the strong interactions, give contributions of the form of equation (15). Symbolically, we thus have:

$$(\Delta M)_\infty \sim \alpha \int_{Q_o^2}^{\infty} \frac{dQ^2}{Q^2} \frac{I_E}{(\ell n\, Q^2/\Lambda^2)p} + \alpha \int_{Q^2}^{\infty} \frac{dQ^2}{(Q^2+M_Z^2)} \frac{I_W}{(\ell n\, Q^2/\Lambda^2)p} \tag{17}$$

where the first term is the e.m. contribution already considered in equation (15), and the second term is the corresponding weak contribution arising from Z exchange; to these must be added possible W± and Higgs terms. The coefficients I_E and I_W are the relative weightings which must be calculated from the weak and e.m. charge structure of the model. It is easy to see that

$$(\Delta M)_\infty \sim \alpha \left[\frac{I_E}{(\ell n Q_o^2/\Lambda^2)^{p-1}} + \frac{I_W}{(\ell n M_Z^2/\Lambda^2)^{p-1}}\right] \quad , \tag{18}$$

showing explicitly that the weak and e.m. quantitatively contribute roughly equally to this piece of ΔM, even though $M_Z^2 \gg Q_o^2$. Thus, in a unified theory with natural symmetry, the presence of the weak interaction not only ensures a convergent result but, by virtue of the asymptotic freedom of the strong interactions, can conspire to give a quantitative contribution comparable to the e.m. Although the standard Weinberg-Salam model does not exhibit a natural symmetry, it is amusing to use it to calculate I_E and I_W. One finds, for the I=1 piece, that

$$I_E \propto + (\Sigma Q_f^2) \text{ and } I_F \propto -(\Sigma Q_f^2)(1-2\sin^2\theta_W) \quad , \tag{19}$$

showing that, for $\sin^2\theta_W \sim 1/2$, the weak contribution enters with the opposite sign to the e.m. In other words, the weak interactions tend to make the neutron heavier than the proton in contrast to the e.m. Whether this is the origin of the sign reversal is certainly open to question, especially since the dynamical symmetry breaking effects presumably give a major contribution. Nevertheless, it is of interest that in a unified theory radiative effects can produce important contributions which naively one might ignore.

IV. THE COTTINGHAM FORMULA AND THE BARE MASS

We have shown in the above discussion that, under certain circumstances, it is possible for the n-p mass difference to be finite and that this was intimately related to a unified electroweak theory. We also showed that, in spite of this, the singlet contribution will, in general remain divergent. For example, in the absence of gluons, one can expect a typical contribution of the form

$$\Delta M_f \sim \alpha \, Q_f^2 \int^\infty dt \; M_f \tag{20}$$

leading to the standard QED type of logarithmic divergence $\sim \alpha Q_f^2 M_f \ell n \, \Lambda$. Now, just as one can define a running coupling constant $g(Q^2)$, so, by treating the mass term $M\bar{\psi}\psi$ as an additional interaction term in the Lagrangian, one can define a running mass $M_f(Q^2)$.[12] This will satisfy a suitable renormalization group equation, and in QCD, will behave similarly to $g(Q^2)$ and fall logarithmically for large Q^2: e.g., $M_f(Q^2) \to [\ell n \; Q^2/\Lambda^2]^{-p'} = t^{-p'}$ where $p' = a_2' \, / \, 2b$. A standard calculation leads to $a_2' = 1/2\pi^2$. Now suppose that instead of the "fixed" mass parameter M_f (the current quark mass) one considered the e.m. contribution to the running mass $M_f(p^2)$. Then one might conjecture that it, too, satisfies an equation analogous to (20). i.e.,

$$\Delta M_f(p^2) \sim \alpha Q_f^2 \int_{p^2}^\infty dt \; M_f(t) \quad , \tag{21}$$

which converges for $p' > 1$. This leads to the constraint $n_f > 21/2$ or $n_f \geqslant 11$. <u>If</u> this argument is indeed valid, it leads to a remarkable result, for, superficially, it says that the e.m. contribution to the self-mass itself is convergent provided there is a sufficient number of flavors. Such a viewpoint has recently been taken by Brodsky, et al.,[13] who justified an equation analogous to (21) using the Schwinger-Dyson equations for the self-mass $\Sigma(p)$. They claim that if one perturbs these equations with lowest order electromagnetism, then in the large p^2 limit, equation (21) follows.

This, of course, raises some delicate and interesting

questions; in particular, how does their result tally with ours which was based on the OPE and which generally "proved" that there is always a logarithmic divergence. In collaboration with Kiskis (who has done most of the work), I have been reviewing this question and in the remaining paragraphs will summarize some of our thoughts on this problem.[14] One possibility is that our starting point, namely the Cottingham formula, is wrong; after all, its structure automatically assumes that the photon integration is performed last and this may not be valid. If the theory were completely finite, then orders of integration would, of course, be irrelevant. However, there are infinities in the loop integrations for both the photon and for the QCD gluon contributions and, for overlapping divergences, in particular, there would be a subtle problem of ordering when constructing counterterms in the renormalization procedure. Symbolically, this can be stated as follows: let Λ_1 be a QED regulator and Λ_2 that for QCD, then a finite perturbative calculation of ΔM can be expressed as $\Delta M\,[e_o, g(\Lambda_1,\Lambda_2), M(\Lambda_1,\Lambda_2),....] \equiv \Delta(\Lambda_1,\Lambda_2)$ where a suitable subtraction scheme has been performed in order to define masses and coupling constants and only lowest order in e_o is implied. To define the physical quantities one must take the simultaneous limits Λ_1 and $\Lambda_2 \to \infty$. A priori, it is not obvious that the answer is independent of the order of taking these two limits, especially since the two interactions are being treated on separate footings. Indeed, the "correct" order presumably is determined by the physics; in order to calculate the corrections to the hadronic mass (m_f) one must first define the unperturbed quantity itself which implies that all renormalizations must first be performed within the hadronic sector before turning on the perturbation. In other words, the "physically relevant" ΔM (to lowest order in e_o^2) is determined operationally by first taking $\Lambda_2 \to \infty$ in order to define an unperturbed M_f, followed by $\Lambda_1 \to \infty$. This immediately tells us that it is indeed the photon integration that must be performed last and, therefore, that the Cottingham formula is, indeed, correct.

The above discussion sheds light on what one means by "correct," namely, that it is the physically relevant quantity. In an unconfined theory, therefore, the Cottingham formula calculates the correction to the mass defined, for example, by the parameter occuring in the mass-shell relationship for a noninteracting particle. For a confined particle, the mass of physical interest has to be defined at some suitable infrared scale. This is in sharp contrast to the quantity calculated by Brodsky, et al.;[13] as already intimated, they have calculated ΔM in the far ultra-violet region where for QCD m itself vanishes.[12] Their quantity, therefore, has less physical relevance being, loosely speaking, the correction to the (vanishing) bare mass and, therefore, nonphysical. In giving this discussion, we have tried to relate two apparently different calculations of ΔM and have concluded that they give different results because they actually are calculating different quantities. There are, in fact, other ambiguities which we have not resolved; for example, the above discussion completely ignored the explicit gluon contribution, which, though absent in the nonsinglet piece, is certainly present for the self-mass itself. However, this apparently plays no role in the calculation of Brodsky, et al.

REFERENCES

1. For a review see A. Zee, Phys. Rep. 3C, 129 (1972); this contains an extensive list of references.
2. See, e.g., S. Weinberg in the Festschrift for I.I. Rabi, ed. L. Motz (N.Y. Academy of Sciences, N.Y. 1977); P. Langacker and H. Pagels, DESY Preprint 78/33 (1978); D.J. Gross, S.B. Treiman and F. Wilzcek Princeton Univ. preprint (Jan. 1979).
3. S. Weinberg, Phys. Rev. Lett. 29, 388 (1972).

4. See, e.g., J. Ellis, lectures at XXIX Session of Les Houches, 1976 (North-Holland, Amsterdam, 1977).

5. G.B. West, lectures at the Erice Highly Specialized Seminars, March 1979 (Plenum Press, to be published).

6. J.C. Collins, Nucl. Phys. B149, 90 (1979).

7. W.N. Cottingham, Ann. Phys. (N.Y.) 25, 424 (1963).

8. H. Harari, Phys. Rev. Lett. 17, 1303 (1966).

9. R. Jackiw, R. van Royen and G.B. West, Phys. Rev. D2, 2473 (1970).

10. J.D. Bjorken, Phys. Rev. 148, 1467 (1966).

11. J.D. Bjorken, Phys. Rev. 179, 1547 (1969).

12. H. Georgi and H.D. Politzer, Phys. Rev. D14, 1829 (1976).

13. S.J. Brodsky, G.F. de Teramond and I.A. Schmidt, Phys. Rev. Lett. 44, 557 (1980).

14. J. Kiskis and G.B. West (to be published).

TESTING STRONG AND WEAK GAUGE THEORIES IN QUARKONIUM DECAYS

Boris Kayser

Physics Division, National Science Foundation

Washington, D.C. 20550

We believe we may have discovered the right gauge theories of both the weak and the strong interactions. However, there is a whole class of processes in which the predictions of these gauge theories have hardly been tested at all. These are the flavor-preserving nonleptonic weak interactions, such as the weak contribution to proton-proton scattering. Being purely hadronic, these processes are weak interactions among quarks, modified by the QCD strong interactions which bind those quarks into hadrons. Thus, they are very difficult to calculate. Being flavor-preserving, these weak interactions have to compete with strong and electromagnetic interactions. Thus, they lead in general to very tiny effects which are almost impossible to see. Consequently, not very much is known about them, although we do have some information from the experiments on parity-violation in nuclei.

I would like to focus here on the flavor-preserving nonleptonic weak interactions of the heavy quarks (charmed, bottom, etc.), about which interactions, needless to say, absolutely nothing is known. However, my collaborators Ephraim Fischbach, Stephen Wolfram, and I believe that the situation is rather more hopeful in the case of the heavy quarks than in that of the light ones, and I would like to try to convince you that there are some heavy quark experiments worth thinking about. In particular, we would like to suggest that the way to study the flavor-preserving non-leptonic weak interactions of the heavy quarks is to look for parity-violation in the decay to ordinary hadrons of the quarkonium states, such as the Ψ/J and the Υ.[1] In Υ decay, for example, you could study the weak transition $b\bar{b} \to u\bar{u}$. This transition, of course, has both the charged- and neutral-current pieces illustrated in Fig. 1.

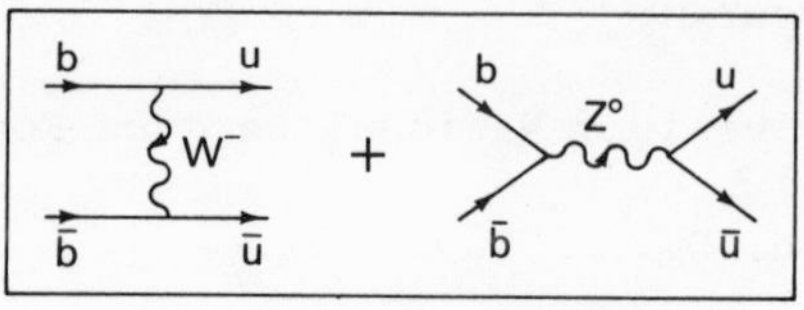

Fig.1. The charged-current (W-exchange) and neutral-current (Z-exchange) contributions to the weak transition $b\bar{b} \to u\bar{u}$.

Before we talk about the kinds of parity-violating signals one can look for, let me tell you how big these effects are expected to be, so you will know whether we are talking about possible physics or science fiction. Let ζ denote generically all the quarkonium states: Ψ/J, Υ, toponium, etc. Parity-violation in hadronic ζ decay, arising from interference between the weak decay amplitude A_w and the strong decay amplitude A_s, will be of order $|A_w/A_s|$.

One estimate of this ratio is $|A_w/A_s| \approx [\Gamma^\zeta_w/\Gamma^\zeta_s]^{\frac{1}{2}}$, where Γ^ζ_w and Γ^ζ_s are, respectively, the weak and strong decay widths of quarkonium. From the standard three-gluon emission picture, the strong decay width is[2]

$$\Gamma^s_s \sim \alpha_s^3 \frac{|\Psi_\zeta(0)|^2}{M_\zeta^2} \quad , \qquad (1)$$

neglecting factors like π. (Here α_s is the strong coupling constant, $\Psi_\zeta(0)$ the quark-antiquark bound state wave function at the origin, and M_ζ the ζ mass.) The weak decay width can be gotten by multiplying the electromagnetic decay width[3]

$$\Gamma(\zeta \to e^+e^-) \sim \alpha^2 \frac{|\Psi_\zeta(0)|^2}{M_\zeta^2} \tag{2}$$

by (M_ζ^2/M_Z^2), thereby replacing the photon propagator in $\zeta \to e^+e^-$ by a Z^o propagator. (Here α is the fine structure constant, and M_Z is the Z^o mass.) Making due allowance for numerical factors which have been omitted from Eqs. (1) and (2), and for some degree of nonleptonic enhancement, we then find that

$$\left|\frac{A_w}{A_s}\right| \approx \left[\frac{\Gamma_w^\zeta}{\Gamma_s^\zeta}\right]^{\frac{1}{2}} \approx \frac{M_\zeta^2}{M_Z^2} \quad . \tag{3}$$

Other crude calculations of the size of parity-violation in quarkonium decay confirm this estimate.[1] In hadronic decays which are electromagnetic rather than strong (because they violate isospin, for example), parity-violation will come from weak-electromagnetic interference. Obviously, in this case we expect a violation of order

$$\left|\frac{A_w}{A_{EM}}\right| \approx \frac{M_\zeta^2}{M_Z^2} \quad , \tag{4}$$

since A_w differs from A_{EM} mainly by having a Z^o (or W) propagator in place of one for a photon. We see that here the parity-violating weak effect is of the same order as when the decay is strong.

Using M_Z = 90 GeV and the actual masses of the quarkonium states, we find from Eq. (3) that in Ψ/J decay, parity-violation will be about a tenth of a percent, in Υ decay about one percent, and in toponium decay at least ten percent, with a precise value that depends on how heavy toponium actually turns out to be. You will notice that parity-violation in quarkonium decay is much bigger than in nuclei, where it is typically of order 10^{-6} or 10^{-7}.

This comes about for two reasons. The first is that, for strong amplitudes, the quarkonium decay amplitudes are anomalously small, as we know from the fact that the quarkonium states are anomalously long-lived. Thus, the weak decay amplitudes have a chance to stick out. The second reason is that the weak interactions grow in relative strength with increasing energy, and the resulting weak effects grow as M_ζ^2. Now, some vital statistics. To see an effect of order ε requires more than $1/\varepsilon^2$ decays. So, for example, in Ψ/J decay, where the effects are of order 10^{-3}, if you are looking in particular at the decay mode $\Psi/J \to \rho\pi$, you will need at least 10^6 $\rho\pi$ decays. Since the branching ratio for $\Psi/J \to \rho\pi$ is 10^{-2}, that means you are going to need 10^8, or, to be more conservative, 10^9 Ψ/J particles. At a "psion-factory" with a luminosity of $L = 10^{32}/\text{cm}^2\text{sec}$ and a beam energy resolution of 0.5 Mev, a billion Ψ/J particles would be accumulated in 15 days. As another example, consider the inclusive decay $\Upsilon \to \pi^+ + \text{anything}$. In Υ decay, parity-violation is expected to be of order 10^{-2}, so you will need at least 10^4 decays which yield a π^+. Since, roughly speaking, every Υ decay yields a pion, a sample of 10^5 Υ particles should suffice. At CESR, if something close to the design luminosity ($0.6 \times 10^{32}/\text{cm}^2\text{sec}$ at the Υ mass) will be achieved when the machine gets fully tuned up, 10^5 Υ particles will be produced in 14 hours. At this meeting we heard two days ago that the luminosity of CESR is currently 1/30 of the design value, but we were promised an additional factor of three essentially for free.[4] Thus, the luminosity is shortly going to be ten percent of its design value. Even with that, the required 10^5 Υ particles would be produced in 140 hours, which is not bad.

Now, what kinds of parity-violating signals does one expect? What is it that one looks for? There are basically two kinds of signals. In the first, illustrated in Fig. 2a, you use a sample of polarized Ψ/J particles with spin $\vec{s}_{\Psi/J}$ and study, for example, a two-body decay mode involving two different particles such as

ρ and π. You look to see whether more pions go up, relative to the spin $\vec{s}_{\Psi/J}$, than down. In other words, you look for a pseudo-scalar correlation $\vec{p}_{\pi}\cdot\vec{s}_{\Psi/J}$ between the momentum of the π and the spin of the Ψ/J. Such a correlation violates parity. In fact, it represents the same kind of parity-violation as that originally seen in ^{60}Co decay.

Alternatively, you can start with unpolarized Ψ/J particles and look for a nonvanishing helicity of a decay fragment with spin. For instance, in the decay $\Psi/J \rightarrow \Lambda\bar{\Lambda}$ illustrated in Fig. 2b, you can look for a nonvanishing helicity of the Λ. This helicity, a pseudoscalar correlation $\vec{s}_{\Lambda}\cdot\vec{p}_{\Lambda}$ between the Λ spin and its momentum, violates parity, and will reveal itself when the Λ decays in a front-back asymmetric way to $p\pi^-$.

Now, will there be backgrounds to the suggested signals from electromagnetism and the strong interactions? If the physical effects we are talking about really violate parity, then, of course, there cannot be any QED or strong backgrounds. So, do they? The answer is "yes and no." If, as in the first half of Fig. 3a, you really started with polarized Ψ/J particles, and saw a front-back asymmetry in $\pi\rho$ decay, this asymmetry would indeed be parity-violating. But that is not what you really do. To study Ψ/J decays, you first make the psions by colliding e^+ and e^- beams. To make polarized psions, you have to longitudinally polarize the beams, as illustrated in the second half of Fig. 3a. Thus, what you really study is not a correlation between the momentum of the π and the spin of the Ψ/J, but between that momentum and the spin of, say, the electron. Unfortunately, since the spin of the electron is parallel to its momentum, there is no way to tell experimentally between a correlation between the momentum of the π and the spin of the electron, which is a pseudoscalar and does violate parity, and one involving the momentum of the π and the momentum of the electron, which is a scalar and does not violate parity. Consequently, the signal you actually look for is not necessarily a

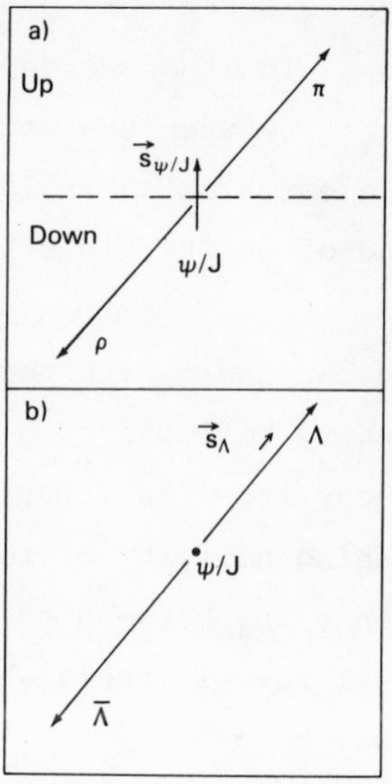

Fig. 2. a) Parity-violating up-down asymmetry of decay pions relative to the spin $\vec{s}_{\psi/J}$ of polarized parent ψ/J particles. b) Parity-violating nonzero helicity of decay lambdas from unpolarized ψ/J particles.

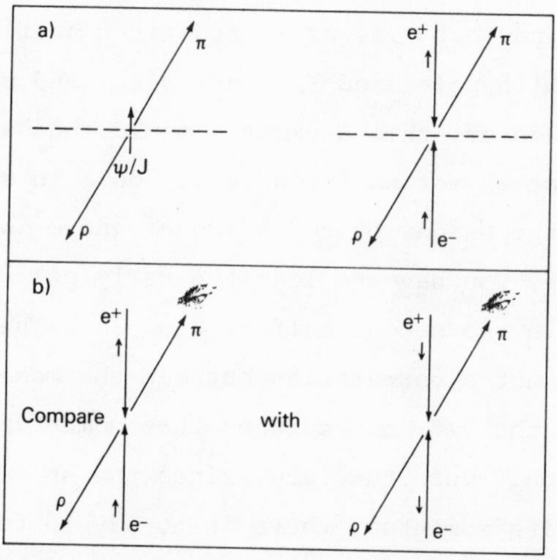

Fig. 3. a) Hypothetical production of $\pi\rho$ from a polarized ψ/J initial state, and corresponding experimentally practical $\pi\rho$ production starting with longitudinally polarized colliding e^+ and e^- beams. The small arrows next to the lines for e^+ and e^- represent their spins. b) Comparison of pion counting rates in a fixed direction ("forward") for two directions of beam spin. Any difference is unambigously parity-violating.

parity-violation, and it can indeed have an electromagnetic background. This background, coming from a diagram with one extra photon relative to the dominant diagram for $e^+e^- \to \Psi/J \to \rho\pi$, is expected to be of order α. Now $\alpha = 1/137$ is bigger than the expected parity-violation, of order 10^{-3}, in Ψ/J decay. Thus, this would be a serious problem, were there not an elegant way to get rid of it. Namely, if you want to distinguish between a correlation that involves the spin of the electron and one that involves its momentum, all you have to do is reverse the electron spin without changing its momentum. If the counting rate changes, then the correlation obviously involves the electron spin, and not its momentum. So, rather than comparing the number of pions which go forward with those which go backward for fixed beam polarizations, fix your detector in the forward direction and, as depicted in Fig. 3b, see if the counting rate changes when you reverse the beam spin. If it does, you are looking at a bona fide parity-violating signal, which cannot receive any contributions from electromagnetic or strong interactions.

That takes care of parity-conserving backgrounds. Now, are there parity-violating backgrounds? Well, yes, there is one. We have been talking about weak Ψ/J decay. However, you can also have a parity-violation coming from weak Ψ/J production followed by strong Ψ/J decay. In other words, replace the photon which normally produces the Ψ/J by a Z^o, as in the top diagram of Fig. 4. To produce a parity-violation, the Z^o must couple to the axial current of the electron and the vector current of the Ψ/J. It cannot be the other way around, because the Ψ/J is a vector particle, and does not have an axial vector current. Now, for fixed beam spins, the Z^o diagram will not lead to a front-back asymmetry. All it does here is contribute to the number of Ψ/J particles which are produced. These Z^o-produced psions decay mainly via the strong interactions, hence in a front-back symmetric way, and no asymmetry results. However, we just learned that to avoid electromagnetic backgrounds, one should not measure the front-back asymmetry for fixed beam spins. Rather, one should fix the detector in the forward direction, and measure the change in counting rate when the beam spins are reversed. Now, the interference between the Z^o diagram and the dominant photon diagram (the bottom diagram in Fig. 4) leads to unequal numbers of Ψ/J particles for given beam intensity in the two beam spin states. Thus, the Z^o production diagram results in unequal forward pion counting rates for the two beam spin states because it leads to unequal numbers of decaying Ψ/J particles in the two cases.

Just as the electromagnetic background can easily be eliminated experimentally, so this weak production background can trivially be dealt with theoretically once someone (Glashow, Salam, and Weinberg, for example) gives you the weak couplings of the Z^o to the axial vector current of the electron and to the vector current of the charmed quark. The weak production background comes from the ratio of the two diagrams in Fig. 4. That ratio involves an unknown but common strong Ψ/J decay amplitude S, which cancels out. It further involves the matrix element $<0|V_\lambda(c)|\Psi/J>$ of the vector current $V_\lambda(c)$ of the charmed quark, both in the Z^o and in the photon diagram. Thus, this matrix element also cancels out. Hence, if you take the Z^o couplings from the very successful

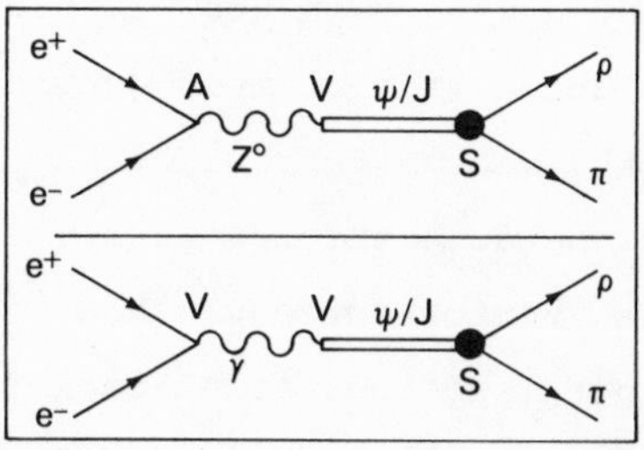

Fig. 4. Diagrams whose ratio gives the parity-violating background from weak Ψ/J production. The symbols V and A indicate vector and axial vector coupling, respectively. The blob labelled S represents the strong decay amplitude for $\Psi/J \to \rho\pi$.

Glashow-Salam-Weinberg model, it is trivial to calculate the ratio of Z^o and γ diagrams, and subtract it from the observed weak signal. The remaining parity-violating signal is from weak Ψ/J decay, and is what we want to study.

To illustrate what physics one can learn from quarkonium decay, let us consider the case of the Ψ/J. What will we learn about the weak Hamiltonian involving the charmed quark by observing parity-violation in hadronic Ψ/J decay? Let us assume that the weak Hamiltonian has a V,A current-current form, with charged- and neutral-current terms. This means that we neglect any Higgs contributions. Let us assume further that there are no flavor-changing neutral currents. In other words, if the $c\bar{c}$ system of which the Ψ/J is made makes a transition to a $u\bar{u}$ pair, it does not do so in the manner shown in Fig. 5. For the neutral-current terms that are left (the flavor-preserving ones, which are the only ones most people assume to be present), there is a simplification. To see this, consider the diagrams in Fig. 6. In the diagrams of Figs. 6a and 6b, the $c\bar{c}$ system which comprises the Ψ/J makes a weak transition to a light quark pair. In Fig. 6a, this weak transition involves W^+ exchange, and in Fig. 6b, Z^o exchange. By contrast, in the diagram of Fig. 6c, the $c\bar{c}$ system makes a Zweig-suppressed strong transition to a light quark pair, and there is a weak interaction into the bargain. This diagram is not only weak, as are the other two, but involves a Zweig-suppressed strong amplitude in addition. Numerically, the Zweig-supressed strong amplitude is some number much smaller than one. Thus, diagrams of this sort are expected to be much smaller than those of Figs. 6a and 6b, and we shall neglect them. To be sure, we do not know exactly how much smaller they are. However, various estimates we have made suggest that they are at least one hundred times smaller in amplitude, so we do believe that they are negligible. Now, if the Z^o contributes to Ψ/J decay only through diagrams such as that in Fig. 6b, then its parity-odd contribution

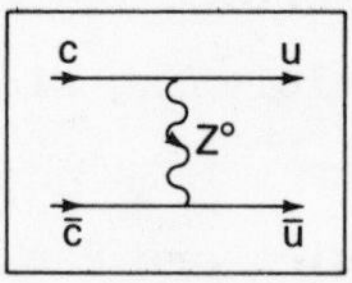

Fig. 5. A presumably absent contribution, involving flavor-changing neutral currents, to the weak transition $c\bar{c} \to u\bar{u}$.

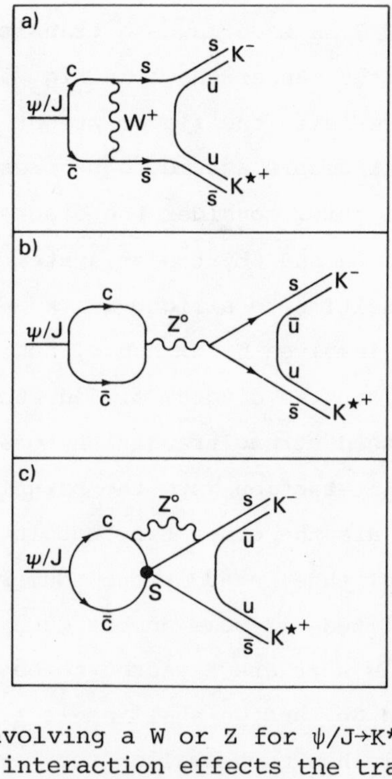

Fig. 6. Diagrams involving a W or Z for $\psi/J \to K^{*+}K^-$. In a) and b), a weak interaction effects the transition from the $c\bar{c}$ system to light quarks. In c), a Zweig-suppressed strong interaction, represented by the blob S, effects this transition, and the diagram contains a weak interaction in addition.

will involve only the vector current of the charmed quark and the axial vector current of the light quarks. No term of the form A (charmed quark) × V (light quarks) will occur, because, as we noted earlier, the Ψ/J is a vector and does not have an axial vector current. Now, in some models, including that of Glashow, Salam, and Weinberg, the light-quark axial current is a pure $\Delta I = 1$ operator so far as u and d quarks are concerned. Thus, the simplification of neglecting diagrams such as that of Fig. 6c leads to a ΔI selection rule.

To summarize, the parity-violating weak Hamiltonian which one can probe by looking for parity-violation in Ψ/J decay to ordinary hadrons has the form

$$\mathcal{H}_{WK}^{p\text{-Viol}} = \frac{G_F}{\sqrt{2}} \{(\bar{c}\gamma_\lambda c)(a_u\bar{u}\gamma_\lambda\gamma_5 u + a_d\bar{d}\gamma_\lambda\gamma_5 d + a_s\bar{s}\gamma_\lambda\gamma_5 s)$$

$$+ b_d[(\bar{c}\gamma_\lambda d)(\bar{d}\gamma_\lambda\gamma_5 c) + (\bar{c}\gamma_\lambda\gamma_5 d)(\bar{d}\gamma_\lambda c)] \tag{5}$$

$$+ b_s[(\bar{c}\gamma_\lambda s)(\bar{s}\gamma_\lambda\gamma_5 c) + (\bar{c}\gamma_\lambda\gamma_5 s)(\bar{s}\gamma_\lambda c)]\} \ .$$

This expression, in which G_F is the Fermi coupling constant, represents the weak interaction among free quarks, uncorrected for QCD effects. In it, the top line is the contribution from the Z^o and involves, as we have said, only the vector current of the charmed quark, and the axial currents of the various light quarks. Three coefficients, $a_{u,d,s}$ enter. The succeeding lines are the charged-current terms connecting c either to d or to s, with two more coefficients, $b_{d,s}$. Altogether there are five coefficients, which in principle are unknown.

Now, what do we already know about these constants $a_{u,d,s}$ and $b_{d,s}$? A die-hard would tell you that we know nothing. He would point out that there may be many $W^\pm$ and Z^o particles. If some W particles contribute to the reactions which have been studied so far, while others contribute to Ψ/J decay, then it is impossible to predict weak effects in the latter from what we have seen in the former. However, experiment supports the standard, one-W, one-Z, SU(2) × U(1) model, including the fermion assignments

$$\begin{pmatrix} u \\ d_c \end{pmatrix}_L , \quad u_R , \quad d_R , \quad \begin{pmatrix} c \\ s_c \end{pmatrix}_L . \tag{6}$$

Here, d_c and s_c denote the Cabibbo-rotated down and strange quarks. Experiment does not fix the remaining quark assignments, and at the moment still allows the right-handed charmed quark to be either in a singlet or a doublet. If it is in a doublet, the partner may be some heavy quark γ, or else this γ mixed with a little bit of the strange quark. The partner cannot be the strange quark itself, or the down quark, or some mixture of those two for reasons that will be explained in Ref. 1. Also, the assignment of the right-handed strange quark is unknown. Given what we do know, it can be shown[1] that the charged-current couplings $b_{d,s}$ are almost completely fixed. The neutral-current coupling a_s between the c quark and the s quark is uncertain, and the neutral-current couplings $a_{u,d}$ between the c and u and d have one of two sets of values, depending on whether the right-handed c quark is in a singlet or in a doublet. When you go from the singlet to the doublet case, the coefficients $a_{u,d}$ quadruple. Now, if we require that neutral currents conserve flavor naturally, then the experimental observation that u_R and d_R are singlets implies that c_R and S_R are as well.[5] Then everything, including the parity-violating coefficients $a_{u,d,s}$ and $b_{d,s}$, is fixed. So far, we have been talking about the weak Hamiltonian for free quarks. Of course, the parity-violating effects one studies will measure, not H_{WK} itself, but its matrix elements $\langle \text{Hadrons} | H_{WK} | \zeta \rangle$ between a quarkonium state and various final hadrons. From such things as the $\Delta I = 1/2$ rule, we know very well that the patterns one sees among the terms of a free-quark Hamiltonian can be very drastically modified by the strong interactions when one is actually looking at the matrix elements of that Hamiltonian. Thus, the parity-violating weak effects which will measure the matrix elements of H_{WK} will test both our understanding of how quantum chromodynamics modifies the weak interactions among free quarks, and our theory of those weak interactions themselves.

We believe that the nonleptonic weak amplitudes for quarkonium decay can be calculated much more easily than the nonleptonic amplitudes that people have worried about before, such as those for K decay, D decay, and hyperon decay. We think that there are some tricks that are special to quarkonium decay that will help a great deal. However, we have not actually calculated these amplitudes yet, and we are mindful of the fact that nonleptonic weak amplitudes have been extremely hard to calculate in the past. So, at this point I really should not say more about this, except to express the fact that we are very optimistic.

What weak effects does one expect, without detailed calculations, in specific illustrative decay modes? Naive quark diagrams should be a rough guide to that. To further convince you of this, let me give you a dishonest argument: It is interesting that the D^o and D^+ lifetimes τ are not equal, as naive free quark diagrams would have required. However, they may not be very different. We learned at this meeting that Fermilab experiment E531 finds[6] that $\tau_{D^o}/\tau_{D^+} < 1/6$, and DELCO finds[7] that $\tau_{D^o}/\tau_{D^+} < \frac{1}{4}$. However, this ratio could, for example, be 1/10, which would mean that the ratio of amplitudes is one to three. And that is not all that different from one to one. You would not be off by orders of magnitude if you just used the naive quark diagram estimate. As another illustration, we note that, while the branching ratios for the Cabibbo-suppressed D decays do not have their free quark diagram values, they are not very different from those values either. The quark diagrams predict

$$\frac{\Gamma(D^o \to \pi^+\pi^-)}{\Gamma(D^o \to \pi^+K^-)} = 0.05 \quad ,$$

and

$$\frac{\Gamma(D^o \to K^+K^-)}{\Gamma(D^o \to \pi^+K^-)} = 0.05 \quad .$$

Experimentally,[8]

$$\frac{\Gamma(D^o \to \pi^+\pi^-)}{\Gamma(D^o \to \pi^+K^-)} = 0.033 \pm 0.015 \; ,$$

and

$$\frac{\Gamma(D^o \to K^+K^-)}{\Gamma(D^o \to \pi^+K^-)} = 0.113 \pm 0.030 \; .$$

This disagreement between theory and experiment is not large, and implies that the ratios of decay amplitudes are quite close indeed to the quark diagram predictions. Thus, without further apology, I would like to consider what one expects to see in some illustrative quarkonium decays using naive quark diagrams as a guide. First, consider $\Psi/J \to \rho\pi$. Here the final state can have I=0,1, or 2. However, one expects to see parity-violating weak effects only in the I=1 final state. (Since the Ψ/J has I=0, this means one expects to see only $\Delta I=1$ weak effects.) The reason is that to violate parity while conserving CP, the weak Hamiltonian must also violate C; that is, it must be odd under C. Now, the Ψ/J is C-odd, so if you act on it with a C-odd Hamiltonian, you will produce a C-even final state. For $\rho\pi$, the C-even state is the I=1 state.

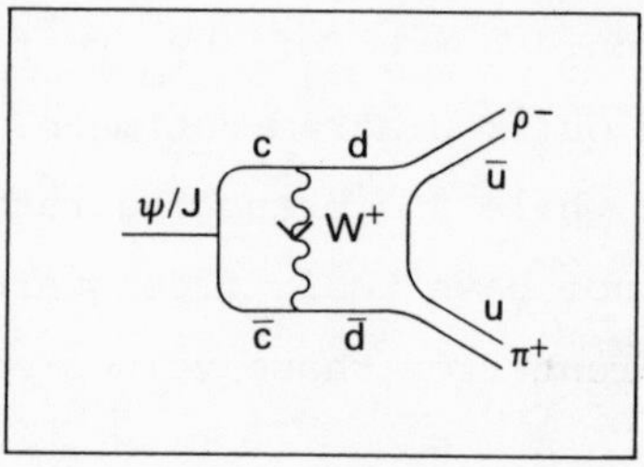

Fig. 7. The charged-current contribution to $\Psi/J \to \rho\pi$.

Now, looking first at the charged-current contribution, shown in Fig. 7, we see that the $c\bar{c}$ system can make a transition to a $d\bar{d}$ pair, which then is dressed to make the final ρ and π. However, the transition to the $d\bar{d}$ pair is Cabibbo-suppressed, at least according to the GIM-charm hypothesis. A transition to an $s\bar{s}$ pair would not be suppressed, but such a transition can play no role here because there are no strange quarks in a ρ or π. Thus, the charged-current contribution is expected to be small, compared to G_F, and the neutral-current contribution should stick out. The situation here is analogous to what one has in some nuclear parity-violating experiments, where by carefully choosing the nuclear states and their isospins, one tries to arrange a situation in which the neutral-current contribution will dominate. Now in our case, the neutral-current contribution will, as we said, involve the vector current of the charmed quark and the axial current of the u and d quarks. As we mentioned, this axial current is expected to be a $\Delta I=1$ operator, as is required here, so we do expect to see a parity-violation. The exact size of this violation will depend on whether the right-handed charmed quark is in a singlet or in a doublet. In the latter case, the effect will be four times as large as in the former. A second, related decay is $\Psi/J \to K^*\bar{K}$. Here the Cabibbo-favored charged-current transition $c\bar{c} \to s\bar{s}$ can indeed contribute, since there are strange quarks in kaons. However, as illustrated in Fig. 8, the $s\bar{s}$ intermediate state necessarily has $I=0$, because strange quarks do not carry isospin. Hence, the charged-current interaction should contribute only to the $I=0$ final state. (The charged-current transition $c\bar{c} \to d\bar{d}$ is Cabibbo-suppressed.) The neutral-current interaction can contribute to both the $I=0$ and $I=1$ final states, so you might wonder how big it is relative to the charged-current term. We have made a crude estimate of that for the Glashow-Salam-Weinberg model. We picture a polarized Ψ/J particle as a noninteracting $c\bar{c}$ pair, with the c and $\bar{c}$ spins aligned parallel to each other. We allow the $c\bar{c}$

system to decay into light quark-antiquark pairs, involving quarks of any flavor and color. Finally, we assume that the asymmetry between the number of light quarks (rather than antiquarks) produced forward and backward relative to the $c\bar{c}$ spin direction is indicative of the front-back asymmetry one would actually see in the final state kaons. This calculation suggests that the neutral- and charged-current contributions are roughly comparable in size. Since the charged-currents contribute only to $\Delta I=0$ transitions, and the neutral-currents to both $\Delta I=0$ and 1 transitions, this means that, barring an accident, there will be comparably-sized parity-violations in the $\Delta I=0$ and $\Delta I=1$ processes. Now, how do we test whether the violations really are comparable? Let us call the strong decay amplitude S, and the weak $\Delta I=0$ and 1 amplitudes W_0 and W_1. In $\Psi/J \to K^{*0}\bar{K}^0$, the weak effect will be proportional to $\mathcal{R}e\, S^*(W_0+W_1)$. In $\Psi/J \to K^{*+}K^-$, it will be proportional to $\mathcal{R}e\, S^*(W_0-W_1)$. Thus, by comparing the weak effects in the two decay charge modes, we can find W_1/W_0. If this ratio turns out to be 50 to 1 instead of 1 to 1, we might wonder whether we understand the nature of the weak interactions in the first place. Perhaps the most interesting kind of parity-violating signal is one that will have to wait for the future. Consider the decay of a really heavy quarkonium state such as toponium. In particular, consider the inclusive decay toponium $\to D_b^-$ + anything. Here, D_b is the expected b-quark analogue of the charmed meson D. As shown in Fig. 9a, the strong decay of toponium presumably goes via three gluons, one of which fragments to make the D_b^-. However, toponium can also decay electromagnetically, as in Fig. 9b, to a $b\bar{b}$ pair, with the quark fragmenting to make the D_b^-. Finally, there is the weak decay, shown in Fig. 9c. This decay also leads to a $b\bar{b}$ pair, with the quark fragmenting to make the D_b^-. Now, from what we already know at current DESY energies,[9] the three gluon decay of toponium should produce three quite well separated jets. The electromagnetic decay, on the other hand, will lead to two-jet events,

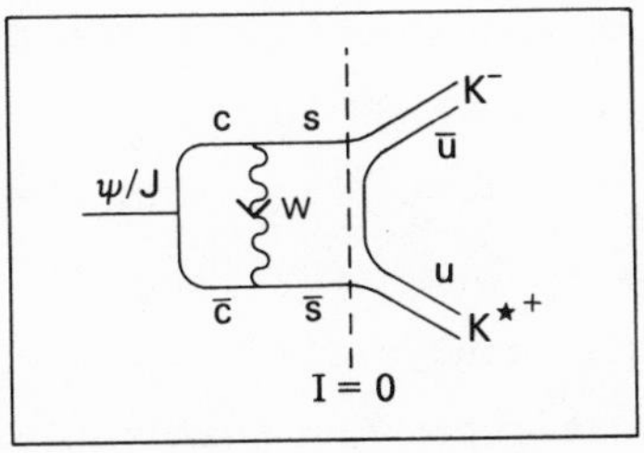

Fig. 8. The charged-current contribution to $\psi/J \to K^*\bar{K}$. Note that the $s\bar{s}$ intermediate state has isospin zero.

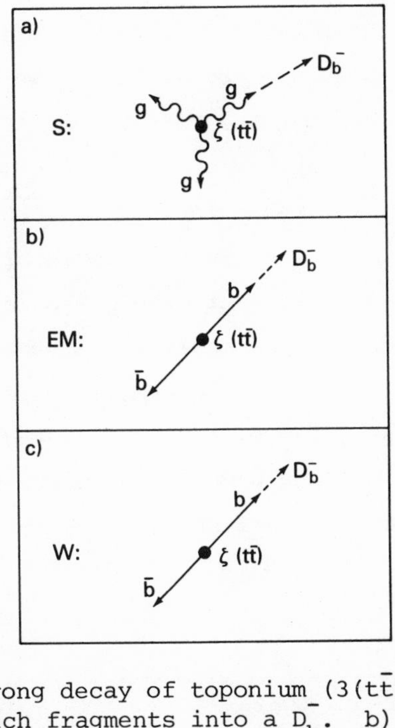

Fig. 9. a) The strong decay of toponium $(3(t\bar{t}))$ via three gluons, one of which fragments into a $D_{\bar{b}}^-$. b) The electromagnetic decay of toponium into a $b\bar{b}$ pair, with the b quark fragmenting into a $D_{\bar{b}}^-$. c) The weak decay of toponium into a $b\bar{b}$ pair, with the b quark fragmenting into a $D_{\bar{b}}^-$.

as will the weak decay. Consequently, because it feeds a different region of phase space than the three gluon amplitude, the weak amplitude will interfere mostly with the electromagnetic one. This circumstance obviously leads to certain computational simplifications. To focus experimentally on the electromagnetically-produced D_b mesons, one should study inclusive D_b^- production in the two-jet decays. This study will teach us about the weak transition $t\bar{t} \to b\bar{b}$. Lest you think that the electromagnetic decays are a miniscule fraction of all toponium decays, I hasten to say that this fraction is not all that small. We estimate that the branching ratio for toponium to decay electromagnetically to $b\bar{b}$ is between two and four percent, depending on how broad toponium actually turns out to be. Furthermore, once the $b\bar{b}$ pair is made, the probability that the b quark will fragment into a D_b^- is obviously quite large. Now, you remember that we expect parity-violation in toponium decay to be bigger than 10%. Thus, one would be seeking in this experiment a striking 10% parity-violation in approximately 1% of all the events. This should be fairly easy to do. To sum up, some of the experiments we are talking about are obviously quite difficult, but others should be rather easy. On the theoretical side, we are hopeful that the nonleptonic weak quarkonium decay amplitudes will be readily calculable. Parity-violation in quarkonium decay will provide a test of the Glashow-Salam-Weinberg model, of QCD, and of the interplay between these two in a class of weak processes about which nothing at all is presently known.

ACKNOWLEDGMENT

I would like to thank Ephraim Fischbach and Stephen Wolfram for a long and fruitful collaboration. The three of us are indebted to many people for helpful conversations, and especially to Alfred Goldhaber, Eugene Golowich, and Barry Holstein.

REFERENCES

1. Preliminary accounts of this suggestion have been given by E. Fischbach in Neutrinos - 78, Proceedings of the International Conference on Neutrino Physics and Astrophysics, Purdue University, edited by E.C. Fowler (Purdue, W. Lafayette, Indiana, 1978), p. 461, and by B. Kayser in Transparencies from the VI International Workshop on Weak Interactions With Very High Energy Beams, 1978, compiled by K.E. Lassila and B.L. Young (Iowa State University, Ames, Iowa, 1978), p. 871. A full discussion of our analysis, including more complete references to the literature, will be published elsewhere.
2. T. Appelquist and H.D. Politzer, Phys. Rev. D12, 1404 (1975).
3. J.D. Jackson, in Weak Interactions at High Energy and the Production of New Particles, proceedings of the SLAC Summer Institute on Particle Physics, 1976, edited by Martha Zipf (SLAC, Stanford, 1976), p. 147.
4. M. Goldberg, talk presented at this conference.
5. S. Glashow and S. Weinberg, Phys. Rev. D15, 1958 (1977).
6. N.W. Reay, talk presented at this conference, and private discussion.
7. G. Donaldson, talk presented at this conference, and private discussion.
8. G. Abrams et al., Phys. Rev. Letters 43, 481 (1979).
9. See, for example, S.L. Wu, talk presented at the Conference on Color, Flavor, and Unification, Irvine, 1979.

SELECTION RULES FOR BARYON NUMBER NONCONSERVATION IN GAUGE MODELS*

R.E. Marshak

Virginia Polytechnic Institute and State University

Blacksburg, Virginia 24061 and

R.N. Mohapatra

City College of City University of New York, New York, N.Y.

ABSTRACT

We discuss the selection rules for baryon number nonconserving processes in the context of various gauge models with partial and complete unification of all elementary particle forces. Three separate cases are discussed: (a) $\Delta(B-L) = 0$, $\Delta(B+L) \neq 0$; (b) $\Delta(B-L) \neq 0$, $\Delta(B+L) = 0$; and (c) $\Delta(B-L) \neq 0$ and $\Delta(B+L) \neq 0$. Observation of $n-\bar{n}$ "oscillation" without proton decay with a lifetime of $\gtrsim 10^{30}$ years would be evidence of "partial unification" with an intermediate mass scale of $\sim 10^{8}-10^{9}$ GeV.

INTRODUCTION

The issue of baryon number nonconservation has been the subject of discussion for a long time[1]. It has attracted a great deal of attention in recent years due to the fact that attempts to unify strong, weak and electromagnetic interactions within a

*Work supported by National Science Foundation

gauge theory framework almost invariably lead to violation of baryon number and hence to the interesting possibility of proton decay.

Once one accepts the possibility of baryon number being only an approximately conserved quantum number, two basic questions arise: (i) the magnitudes of various B-violating amplitudes; and (ii) the kinds of selection rules governing these processes. The present experimental limits[1,2] on the nucleon decay lifetime of $\tau_N \gtrsim 10^{29-30}$ years, imply that the strengths of B-violating processes are very weak indeed. This result interpreted within the gauge theoretical picture of unifying all elementary particle forces is the basis for the current wisdom that there exists a superheavy mass scale $10^{14} - 10^{15}$ GeV in physics. Coming to the second question, particular theoretical schemes[3,4] lead to characteristic selection rules for B-violating processes. However, in the absence of any decisive evidence to believe in any given scheme, it is useful to study this question through a variety of models so as to highlight the various processes that could discriminate not only between various grand unification schemes but also could inform us as to the necessity even to go beyond the level of "partial" unification.

An important observation bearing on the general question of selection rules is due to Weinberg[5], Wilczek and Zee[5]. They note that, if baryon nonconservation is mediated by local four-fermion operators (i.e. no derivative couplings) that respect the $SU(2)_L \times U(1) \times SU(3)_c$ group of the standard low energy electroweak and strong interactions, then B-L is an exact conserved quantum number, permitting decays like $p \to e^+\pi^0$, $n \to e^+\pi^-$, etc. (see Table 1). This selection rule eliminates the electron decay channels as well as nonleptonic channels[6]. In the context of specific grand unification schemes, depending on whether there exists absolutely conserved global B-L symmetry in the theory or not, $\Delta(B-L) \neq 0$ processes may or may not be absent altogether due to higher order effects. These are in general expected to be small. However, under certain circumstances[7], B-L violating decays may be of experimental interest and throw light on the nature of the underlying theory of elementary particles.

Before beginning our detailed discussion, we first note that deviations from the above theorem can arise in the limit of exact $SU(2)_L \times U(1) \times SU(3)_c$ symmetry provided we allow for four-fermion operators with derivative couplings or for local six-fermion operators. Examples are:

$$O^{(4)} = \bar{\ell}_{Lp}\, \gamma_\mu\, q^p_{L,i}\, \bar{q}^c_{L,j} D_\mu\, q_{R,k}\, \varepsilon^{ijk} \quad ,$$

$$O^{(6)} = \varepsilon^{i_1 j_1 k_1}\, \varepsilon^{i_2 j_2 k_2}\, \bar{d}^c_{L,i_1}\, d_{R,i_2}\, \bar{d}^c_{L,j_1}\, d_{R,j_2}\, \bar{u}^c_{L,k_1}\, u_{R,k_2} \,. \quad (1)$$

Here D_μ is the effective color gauge-invariant derivative coupling, p denotes the SU(2) index and i, j, k... denotes color index. The first operator $O^{(4)}$ would lead to decays of the type $n \to e^- \pi^+$ and the second operator $O^{(6)}$ would lead to transitions of the type $n \leftrightarrow \bar{n}$, which we have called "neutron oscillations"[8]. In the rest of this review, we present theoretical schemes where these operators dominate and discuss some of their consequences.

To systematize the discussion, we define three classes of B-violating reactions on the basis of the selection rules obeyed by them (see Table 1): (i) Processes with $\Delta(B-L) = 0$ but $\Delta(B+L) \neq 0$. This includes decays such as $p \to e^+\pi^0$, $n \to e^+\pi^-$, etc. (ii) Processes with $\Delta(B-L) \neq 0$ but $\Delta(B+L) = 0$. The decays of this class are $p \to e^-\pi^+\pi^+$, $n \to e^-\pi^+$, etc. (iii) Processes with $\Delta(B-L) \neq 0$ and $\Delta(B+L) \neq 0$. This includes transitions such as n-$\bar{n}$ ("neutron oscillation") and the Pati-Salam mechanism[6]. We now discuss various gauge models that give rise to new selection rules in B-violation.

Table 1. A summary of the various selection rules in baryon nonconserving processes and their theoretical implications.

					Model Implications if Observed with $\tau_N \sim 10^{30}$-10^{33}yrs.			
B-L	B+L	Process	Present Experimental Limit (yrs)*	Achievable Limit*	SU(5) (minimal)	$SU(2)_L \times SU(2)_R \times SU(4')$ (minimal)	$SU(2)_L \times SU(2)_R \times SU(4')$ or SU(5) with extra Higgs	New Theoretical Idea necessary
Conserved	Violated	$p \to e^+ \pi^0$ $n \to e^+ \pi^-$	10^{29} 10^{29}	10^{33}	yes	no	yes	yes
Violated	Conserved	$p \to e^- \pi^+ \pi^+$ $n \to e^- \pi^+$	10^{30}	10^{33}	no	no	no	yes
Violated	Violated	$n \to \bar{n}$	10^{30}	10^{33}	no	yes	yes	yes

*According to Prof. Reines (private communication), the "present experimental limits" are derived from the South African detector results and the "achievable limits" might be reached with the 10,000 ton Irvine-Michigan-Brookhaven detector now under construction.

LOCAL B-L ELECTROWEAK SYMMETRY AND ITS BREAKDOWN

In a recent note[8], we have pointed out that in the left-right symmetric model of weak and electromagnetic interactions based on the $SU(2)_L \times SU(2)_R \times U(1)$ group[9], the U(1) generator corresponds precisely to a B-L local symmetry. Therefore, breakdown of parity in this model gets related to the breakdown of B-L symmetry[10]. This connection is seen clearly by noting the formula for the electric charge[11], i.e.

$$Q = I_{3L} + I_{3R} + \frac{B-L}{2} \quad . \tag{2}$$

Thus, since $\Delta Q = 0$, at the energy scale where $\Delta I_{3L} \simeq 0$ (i.e. $E \sim 200$ to 300 GeV), Eq. (2) implies that,

$$\Delta I_{3R} \simeq -\frac{1}{2}\,\Delta(B-L) \quad . \tag{3}$$

In a further extension of the quark-lepton symmetry principle, we enlarge the group $SU(2)_L \times SU(2)_R \times U(1)_{B-L} \times SU(3)_c$ to the "partial unification" group $SU(2)_L \times SU(2)_R \times SU(4')$, where B-L is identified as the fourth color[12]. We then establish[8] the existence of $0^{(6)}$ through graphs of the type shown in Fig. 1. A rough estimate of the strength of neutron oscillation can be given:

$$A_{n-\bar{n}} \simeq \frac{\lambda\, h^3 \langle\Delta_R\rangle}{m_\Delta^6} \quad , \tag{4}$$

where λ is the scalar self-coupling, h is the strength of fermion-Yukawa coupling, m_Δ is the mass of the Higgs scalars. Plausible values for these parameters are: $\lambda \simeq \alpha^2$, $h \sim \alpha$, $\langle\Delta_R\rangle \simeq 10^3$GeV. Thus, if we choose, $m_{\Delta_R} \simeq 10^9$GeV, we obtain a lifetime for neutron oscillation[8] $\sim 10^{31}$ years. It is therefore within accessible range of the present experiments searching for proton decay[13] (see Table 1).

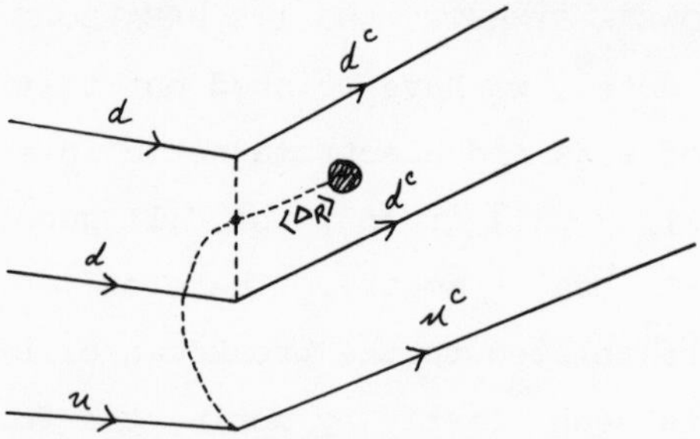

Fig. 1 A typical Feynman graph generating the effective six-fermion operator leading to neutron oscillation. The solid lines correspond to the quarks, the dashed lines denote the "minimal" Higgs scalars and $\langle\Delta_R\rangle$ denotes the strength of the vacuum-breaking of parity symmetry.

Two other important aspects of this model are worth stressing: (i) the minimal version of this model that leads to the n-$\bar{n}$ transition incorporates one Higgs multiplet of type (1,1,15), that breaks SU(4') to $SU(3)_c \times U(1)_{B-L}$, and the flavor-Higgs multiplets transforming as (2,2,1) and ((1,0) + (0,1), 10) under $SU(2)_L \times SU(2)_R \times SU(4')$. In this minimal version, there actually exists a discrete quark symmetry $a \to e^{\frac{i\pi}{3}} q$ which prohibits proton decay. Thus, observation of neutron oscillation with no proton decay in the lifetime range 10^{30-32} yrs. would provide a clear signature for the existence of a partial unification model with an intermediate mass scale of about 10^9 GeV. (ii) The second point we stress is that the above selection rules can be modified if we include an additional Higgs multiplet (0,0,6). Proton decay is then allowed via a diagram of the type shown in Fig. 2. But, unless the new Yukawa and scalar meson couplings involving this extra Higgs boson are very much larger, this decay is quite suppressed. (For example, if $h_6' \sim 1$ and $\lambda_6 \sim 1$, we get $\tau_p \approx 10^{34}$ yrs., if all Higgs boson masses are chosen to be of the same order.) The (B-L)-violating (but (B+L)-conserving) - nucleon decay lifetime (e.g. $p \to e^-\pi^+\pi^+$, $n \to e^-\pi^+$, etc.) in this model is even longer.

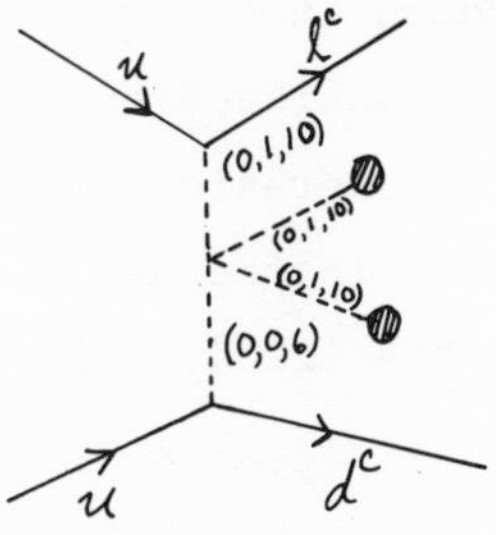

Fig. 2. A typical Feynman diagram giving rise to the (B-L)-conserving proton decay ($p \to e^+\pi^0$).

SU(5) MODEL WITH EXTRA HIGGS MESONS

In the minimal SU(5) model with Higgs mesons belonging to {5} and {24} dimensional representations, there exist global symmetries[7] that keep B-L conservation exact to all orders. However, this selection rule is violated by including[14] a symmetric {15} dimensional representation denoted by $S_{\mu\nu}$. This couples to the fermions belonging to the {5} dim. representation (ψ_μ) of SU(5) as follows:

$$L = h_S \, \psi_\mu^T C^{-1} \psi_\nu \, S_{\mu\nu} \quad . \tag{5}$$

Since $\langle S_{55} \rangle = v \neq 0$ and $v \sim$ GeV, Eq. (5) would give Majorana mass to the neutrinos. Thus $m_\nu < 10$ eV implies $h_S < 10^{-8}$. The Lagrangian in Eq. (5) induces both n-$\bar{n}$ oscillation as well as (B+L)-conserving nucleon decay ($n \to e^-\pi^+$, etc.); however, with $\tau_{n\bar{n}} \gtrsim 10^{30}$ years, $t_{n \to e^-\pi^+} \gtrsim 10^{58}$ years, i.e. the (B+L)-conserving decays would be highly suppressed in this model.

COMMENT ON SO(10) GRAND UNIFICATION

In this section, we would like to remind the reader that the $SU(2)_L \times SU(2)_R \times SU(4')$ model discussed in Section 2 as the simplest embodiment of quark-lepton unification is easily accommodated in an SO(10) grand unified theory.[15,16] The particular pattern of symmetry breakdown that gives rise to a significant n-$\bar{n}$ transition amplitude can be achieved as follows, assuming that the quarks and leptons of each generation are assigned to the 16-dimensional spinor representation in the usual manner[15,16]. We need the Higgs mesons belonging to the {210}, {45}, {126} and {10} dimensional representations. The choice $\langle\phi_{210}\rangle \neq 0$ breaks the symmetry down to $SU(2)_L \times SU(2)_R \times SU(4')$. We assume the next stage of breakdown to arise due to $\langle\phi_{45}\rangle \neq 0$, which leaves $SU(2)_L \times SU(2)_R \times U(1)_{B-L} \times SU(3)_c$ as the surviving symmetry. The breakdown of parity arises from $\langle\phi_{126}\rangle \neq 0$ which causes $\Delta B \neq 0$ and $\Delta L \neq 0$ phenomena like Majorana neutrino and n-$\bar{n}$ oscillation that we have discussed. The final breakdown that leaves only $U(1)_{em} \times SU(3)_c$ is caused by $\langle\phi_{10}\rangle \neq 0$ (several of these multiplets), that set the scale for the low energy weak interaction and also give mass to the quarks and charged leptons (see Fig. 3). We present these remarks in a purely phenomenological spirit[17] in order to indicate that the phenomenon of n-$\bar{n}$ oscillation could also arise in a grand unified scheme, with a unification scale much higher than that of SU(5).

$$SO(10)$$

$$\Big|\quad m_{X,Y} \simeq g\langle\phi_{210}\rangle$$

$$SU(2)_L \times SU(2)_R \times SU(4')$$

$$\Big|\quad m_{PS} \simeq g\langle\phi_{45}\rangle$$

$$SU(2)_L \times SU(2)_R \times U(1)_{B-L} \times SU(3)_c$$

$$\Big|\quad m_{W_R} \simeq g\langle\phi_{126}\rangle$$

$$SU(2)_L \times U(1)_{T_{3R}+\frac{B-L}{2}} \times SU(3)_c$$

$$\Big|\quad m_{W_L} \simeq g\langle\phi_{10}\rangle$$

$$U(1)_{em} \times SU(3)_c$$

Fig. 3. Showing the successive stages in the breakdown of SO(10) symmetry and the grand-Higgs-multiplets responsible for the breakdown at each stage.

SUMMARY

In summary, we wish to point out that study of the various selection rules in B-violating processes can yield significant information concerning the underlying structure of the elementary particles, in particular the quarks and leptons. Table 1 contains a summary of the model predictions as well as of the experimental situation. It should be noted that the predictions of the "grand unification" group SU(5) are characteristic of other "grand unification" groups like SO(10). These predictions differ in a striking way from those of the "partial unification" group $SU(2)_L \times SU(2)_R \times SU(4')$.

ACKNOWLEDGEMENT

We would like to thank L.N. Chang, M. Goldhaber, L. Mo, F. Reines and G. Senjanovic for valuable discussions at various times. Also, we have just learned from Prof. S.L. Glashow that he has studied the "neutron oscillation" and electron modes of nucleon decay within the framework of the extended SU(5) model in two Harvard preprints[18].

REFERENCES AND FOOTNOTES

1. For a review and earlier references, see M. Goldhaber, "Unification of Elementary Forces and Gauge Theories", ed. by D. Cline and F. Mills, Academic Press (1977), p. 531. H.S. Gurr, W.R. Kropp, F. Reines and B.S. Meyer, Phys. Rev. 158, 1321 (1967): F. Reines and M.F. Crouch, Phys. Rev. Lett. 32, 493 (1974).
2. J. Learned, F. Reines and A. Soni, Phys. Rev. Lett. 43, 907 (1979).
3. J.C. Pati and A. Salam, Phys. Rev. Lett. 31, 661 (1973).
4. H. Georgi and S.L. Glashow, Phys. Rev. Lett. 32, 438 (1974).
5. S. Weinberg, Phys. Rev. Lett. 43, 1566 (1979). F. Wilczek and A. Zee, Phys. Rev. Lett. 43, 1571 (1979).

6. In the Pati-Salam model with integer charge quarks, this theorem is not respected since $SU(3)_c$ is only an approximate symmetry at low energies. Since their model has an absolutely conserved B+3L quantum number (called by them fermion number), it allows for decays of the type $p \rightarrow \pi^+ + 3\nu$. For a review and detailed predictions, see J.C. Pati, University of Maryland Technical Report No. 79-066 (1979) (unpublished).

7. The question of B-L violation in the framework of grand unification has been considered by F. Wilczek and A. Zee, University of Pennsylvania Preprint, (1979).

8. R.N. Mohapatra and R.E. Marshak, VPI-HEP-80/1.

9. J.C. Pati and A. Salam, Phys. Rev. D10, 275 (1974). R.N. Mohapatra and J.C. Pati, Phys. Rev. D11, 566, 2558 (1975). G. Senjanovic and R.N. Mohapatra, Phys. Rev. D12, 1502 (1975). For a review, see R.N. Mohapatra, "New Frontiers in High Energy Physics", ed. by A. Perlmutter and Linda Scott, (Plenum, 1978), p. 337.

10. This yields $\Delta L = 2$ processes such as neutrinoless double β decay and an interesting correlation between neutrino mass and the right-handed W boson, i.e. $m_\nu \simeq \frac{m_\ell^2}{g_{m_{W_R}}}$. See R.N. Mohapatra and G. Senjanovic, CCNY-HEP-79/10 (1979).

11. The equivalent of this relation was suggested by A. Gamba, R.E. Marshak and S. Okubo (Proc. Nat. Acad. of Sci.? (1959)) when the consequences of the baryon-lepton symmetry of the weak interaction were first examined. See also paper by R.E. Marshak and R.N. Mohapatra for Maurice Goldhaber Festschrift (New York, 1980).

12. We have used the same notation as J.C. Pati and A. Salam (ref. 9) but it is important to stress that the fourth color in our case is (B-L) not L.

13. For a review of the proposed experiments, see L. Sulak, Proceedings of "Weak Interaction" Workshop at Virginia

Polytechnic Institute (1979).

14. The extended SU(5) model has also been considered in detail by L.N. Chang and N.P. Chang (to be published).

15. H. Fritzsch and P. Minkowski, Ann. of Phys. 93, 193 (1975). H. Georgi, in "Particles and Fields, 1975" (AIP Press, N.Y.).

16. M. Chanowitz, J. Ellis and M.K. Gaillard, Nuc. Phys. B129, 506 (1977). For recent discussions on the subject see H. Georgi and D.V. Nanopoulos, Nuc. Phys. B155, 52 (1979). R.N. Mohapatra and B. Sakita, Phys. Rev. D (to appear). M. Gell-Mann, P. Ramond and R. Slansky - unpublished.

17. This is the subject of a forthcoming paper by R.N. Mohapatra and G. Senjanovic, to appear as a City College Preprint (1980). It is shown in this paper that to get $\sin^2\theta_W(m_{W_L})$ to be about 0.23, while at the same time getting an intermediate mass scale, requires the value $m_{X,Y} \approx 10^{19}$GeV and $m_{W_R} \approx 10^9$GeV, as well as $m_{X,Y} \approx m_{PS}$. In this case, the proton decay mediated by the gauge bosons is completely suppressed leaving n-$\bar{n}$ oscillation as a possible dominant mode of baryon nonconservation.

18. HUTP-79/A029 and HUTP-79/A059. These preprints discuss modifications of the minimal SU(5) model and reach similar conclusions to ours as summarized in Table 1; however, we stress that the new result highlighted by Table 1 is the predicted dominance of the "neutron oscillation" mode in a partial unification model incorporating B-L local symmetry on the electroweak level.

NONLINEAR EFFECTS IN NUCLEAR MATTER

AND SELF-INDUCED TRANSPARENCY*

G. N. Fowler

University of Exeter, England and

R.M. Weiner

University of Marburg, W. Germany

(presented by R.M. Weiner)

ABSTRACT

We suggest that in high energy hadron-nucleus and nucleus-nucleus reactions a partially coherent intense mesonic field is created. Two non-linear models for the propagation of this mesonic field in nuclear matter are considered which both admit solutions. The effect of this could be a selfinduced transparency of nuclear matter. This might explain various experimental observations in hadron-nucleus reactions and leads possibly also to the existence of new metastable states. Experimental consequences of this are discussed.

In hadron collisions with nuclei at high energies the supression of cascading[1] (transparency effect) which is observed is commonly interpreted in terms of a coherent interaction between the projectile and the nuclear matter traversed. In the first

*Work supported in part by NATO, BMFT and GSI Darmstadt.

encounter of the projectile within the nucleus an intense coherent mesonic field is expected to be created. It thus seems plausible that the response of the nuclear medium will be non-linear in the mesonic field.

Furthermore and particularly in the case of heavy ion collisions at high energy one has the added complication that quite apart from possible high meson field intensities the nuclear density is also expected to undergo large fluctuations, implying a significant interaction in the meson field source.

Of course to treat these problems in anything like full generality is out of the question quite apart from the fact that much of the basic interaction required for a solution is unknown. Nevertheless on the basis of models drawn from optics and solid state physics it is possible to make certain qualitative suggestions. In particular, in this paper we apply the formalism used in the description of self-induced transparency (SIT) to high energy nuclear reactions and discuss some possible tests of this approach.

NON-LINEAR EFFECTS IN THE MESONIC FIELDS

The conditions for SIT to develop as discussed by McCall and Hahn[2] are:

i) An intense, coherent (classical) field which interacts with the medium.

ii) The existence of at least two energy levels in the medium - such that the excited level has a lifetime larger than that of the coherent pulse.

If one assumes, as indicated by recent experiments (cf. below), that in the collision of a highly energetic hadron with a nucleon in the nucleus an intense coherent pion field is created, conditions (i) and (ii) could be met in nuclear matter. Among other things, the condition that the pulse should be broad enough in the direction of propagation to overlap more than one nucleon is very similar to that postulated by the hydro-dynamical and coherent tube model. At laboratory energies of 200-400 GeV the levels to be excited in the nuclear medium would be in the 5-10 GeV region where a continuum of resonances might exist. The resonance lifetime in the nucleus (laboratory frame) is enhanced by Lorentz effect. The pion field satisfies the Klein Gordon equation.

$$(m^2 + \Box)\phi = j \quad , \tag{1}$$

where j is the current induced by ϕ and m the pion mass; we neglect meson-meson interactions in this first overview. Within the SIT mechanism the width Γ of the pulse is given by

$$\Gamma = g\,\phi/2 \quad , \tag{2}$$

where g is the pion nucleon coupling constant (we write the pseudo-scalar pion-nucleon isoscalar interaction $g\underline{\phi}\cdot\underline{\tau}\bar{\psi}\gamma_5\psi$).

Consider for ϕ and j a plane wave approximation

$$\phi = \tilde{\phi}(z-vt)e^{i(kz-\omega t)} \quad ; \qquad j = \tilde{j}(z-ve)e^{i(kz-\omega t)} \quad , \tag{3}$$

where $\tilde{\phi}$ and j are slowly varying functions.

z denotes the propagation direction (we treat the problem in one dimension only, which should be satisfactory at high enough energies where the average P_T of secondaries is limited); ω and k are the energy and wave number of the pulse in the medium which are to be determined.

Substitution of (3) into (1) yields

$$-(\omega^2-k_o^2)\tilde{\phi} + 2ik\frac{\partial\tilde{\phi}}{\partial z} + 2i\omega\frac{\partial\tilde{\phi}}{\partial t} = \tilde{j} \; ; \quad k_o = \sqrt{k^2+m^2} \quad . \tag{4}$$

Chosing $\tilde{\phi}$ real, one has

$$-(\omega^2 - k_o^2)\tilde{\phi} = \mathrm{Re}\,\tilde{j} \tag{5}$$

$$- 2k \frac{\partial \tilde{\phi}}{\partial z} k - 2\omega \frac{\partial \tilde{\phi}}{\partial t} = \mathrm{Im}\, \tilde{j} \quad , \tag{6}$$

The nucleon wave function of a two level systems reads

$$|(t)> = a(t)|\psi_o> + b(t)|\psi_1> \quad , \tag{7}$$

where ψ_o and ψ_1, denote the ground and excited level respectively. The current j is given in terms of a, b and $\tilde{\phi}$ by the Bloch equations,

$$\frac{\partial w}{\partial t} = - g \tilde{\phi} w \quad ,$$

$$\frac{\partial v}{\partial t} = g \tilde{\phi} w \quad , \tag{8}$$

$$\frac{\partial u}{\partial t} = 0 \quad ,$$

with

$$g(ab^* + a^*b) = n^{-1} \mathrm{Re} j \equiv gu$$

$$ig(ab^* - a^*b) = n^{-1} \mathrm{Im} j \equiv gv$$

$$bb^* - aa^* \equiv w \ ; \quad aa^* + bb^* = 1 \ . \tag{9}$$

w is the inversion which is the essential factor in the saturation effect necessary for SIT, and n is the number of nucleons per unit volume. For simplicity only one resonance level has been assumed and the resonance condition $\omega = \bar{\omega}$ has been postulated, where ω is the pion energy corresponding to the excited level $\bar{\omega}$. This means that only resonant collisions are considered, which are expected to be the main contributors to nuclear opacity.[4]

It is known from the theory of SIT that these simplifications do not alter the main characteristics of the effect[4]. In particular transparency is also present when the slowly varying approximation is inadequate, which may be a characteristic of the hadron case[5]. The soliton[3] solution of Eq. (5) reads

$$\begin{aligned} w &= \cos\theta \quad , \\ v &= \sin\theta \quad , \\ u &= 0 \quad , \end{aligned} \tag{10}$$

with

$$\phi = g \int_{-\infty}^{t} \tilde{\phi}\, dt \quad , \tag{11}$$

which can be inverted to

$$\frac{\partial\theta}{\partial t} = g\,\tilde{\phi} \quad . \tag{12}$$

Substitution of (12) into (6) yields the sine-Gordon equation,

$$\frac{\partial^2\theta}{\partial\zeta^2} - \frac{\partial^2\theta}{\partial\tau^2} = \frac{n}{2k} \sin \quad , \tag{13}$$

where

$$\zeta = g(t - 2z) \ , \ \tau = gt \ .$$

Eq. (13) admits the solution

$$\tilde{\phi} = g\,\tilde{\phi}_o \operatorname{sech}\left[g\,\frac{\tilde{\phi}_o}{2}\,(t - \frac{z}{v})\right] \quad , \tag{14}$$

where $\tilde{\phi}_o$ is an arbitrary constant. The system of equations (6), (10) and (11) leads to the "area" theorem

$$\frac{d\theta}{dz} \propto - \sin\theta \quad , \tag{15}$$

which implies that the intensity of the pulse is not attenuated in the medium provided θ is a multiple of π for large times. This in turn requires that the pulse is intense, i.e. θ large. On the contrary, in the $\theta \to 0$ limit Eq. (15) yields an attenuated pulse (Lambert's Law). It is interesting to emphasise that the strength of the pulse is determined both by the magnitude of the coupling g and the amplitude of the field $\tilde{\phi}_0$, so that when using hadronic (pion) fields one can afford to have smaller field amplitudes as compared with the optical (electromagnetic) case because of the favourable $g/\sqrt{\alpha}$ ratio, α being the fine structure constant.

In the derivation of the transparency effect an essential element was the coherence of the meson field. This gives a more precise form to the conjecture often made in the literature that the nuclear transparency is due to the long formation time of the secondary hadrons compared with nuclear dimensions. However, transparency can also be produced by incoherent fields ("hole-burning") and if the tests for coherence referred to below fail, this may be a possible approach. In any event both procedures rely on saturation effects in the nuclear medium.

NON-LINEAR EFFECTS IN THE NUCLEAR DENSITY

If now we take the view that saturation of the nuclear medium is unlikely but that high density nuclear matter is created, a different model is required. Since this has not been so extensively explored we present some details here.

We start from a Lagrangian of mesons and meson field sources in a one dimensional approximation[6]

$$L = \frac{1}{2}\left(\frac{\partial\rho}{\partial t}\right)^2 - \frac{1}{2}\left(\frac{\partial\rho}{\partial z}\right)^2 - \frac{1}{2}\,b\rho^2 - \frac{1}{4}\,c\rho^4 + g\rho\phi + L_\pi\,, \qquad (16)$$

where $\rho(z,t)$ is the meson field source, $\phi(z,t)$ the meson field variable, L_π is the free meson field Lagrangian (in which we ignore meson-meson interactions), b is an effective mass of the nucleonic excitation, and c a self-coupling constant of the sources. To fix ideas on can picture the source field as creating or annihilating πN resonances which are allowed propagate through nuclear matter with a $-\rho^4$ interaction term. Since ρ and ϕ are pseudoscalars no odd powers in ρ alone appear. The one dimensionality of the problem is expected to arise from the well-known limitation on the transverse momentum of pions produced in nucleon-nucleon collisions.

The equations of motion are now

$$\frac{\partial^2 \rho}{\partial t^2} - \frac{\partial^2 \rho}{\partial z^2} = -b\,\rho - c\rho^3 + g\phi \;, \tag{17}$$

$$(m^2 + \Box)\,\phi = H \;. \tag{18}$$

We now assume that the fields may be treated as c numbers (coherence property) and that the intensities are high enough for classical behavior to be a reasonable approximation. Finally we write for ρ and ϕ

$$\rho = \tilde{\rho}(z,t)e^{i(kz-\omega t)} \equiv (\tilde{\rho}_R(z,t) + i\,\tilde{\rho}_I(z,t)e^{i(kz-\omega t)} \;, \tag{19}$$

$$g = \tilde{\phi}(z,t)e^{i(kz - \omega t)}$$

where $\tilde{\rho}$ and $\tilde{\phi}$ are slowly varying.

Two different extreme cases may be investigated namely:

(i) $\rho_R >> \rho_I$,

(ii) $\rho_I >> \rho_R$. (cf. references (7), (8) and (9)).

For simplicity we shall discuss only case (i); the other can be treated on similar lines.

Thus we may take(see also references (7) and (8))

$$\tilde{\phi}(z,t) = \frac{\tilde{\rho}_R(z,t)}{k^2+m^2-\omega^2} \tag{20}$$

and substitute in (17). If we write:

$$\tilde{\rho}_R\,(z,t) \equiv \tilde{\rho}_R(\frac{1}{\tau}\,(t - z/v)) \equiv \tilde{\rho}_R(z)\;, \tag{21}$$

we find that (17) becomes:

$$\frac{1}{\tau^2}(1 - \frac{1}{v^2})\,\ddot{\tilde{\rho}}_R + (k^2 - \omega^2 + b - \frac{g^2}{k^2+m^2-\omega^2})\,\tilde{\rho}_R + c\tilde{\rho}_R^{\;3} = 0\,. \tag{22}$$

The solution to (22) is the usual sech pulse soliton

$$\tilde{\rho}_R = \tilde{\rho}_o \text{ sech } (\frac{1}{\tau}\,(t-z/v))\;, \tag{23}$$

where

$$\tau^2 = \frac{2(1-1/v^2)}{c\rho_o^{\;2}}\;, \tag{24}$$

with the dispersion relation

$$k^2 - \omega^2 + c - \frac{g^2}{k^2+m^2-\omega^2} + \frac{c\rho_o^{\;2}}{2} = 0\,. \tag{25}$$

In this approximation v is arbitrary but an additional relation arises in a more exact treatment (references 8, 9 and 10).

A condition for this non-linear excitation is $\tau < \tau_{relax}$, where τ_{relax} is the relaxation time of the nuclear excitation; for high enough ρ_o (nuclear density fluctuation), this should be true. The existence of a soliton solution in $\tilde{\rho}$ leads again to transparency of nuclear matter for pions.

CONCLUSIONS

We now turn to the qualitative predictions to which models of the type we are discussing give rise.

1. If a coherent meson field is indeed responsible for the transparency effect, then we should expect it to manifest itself in the Bose-Einstein correlations of secondary pions as described, e.g., in references 11 and 12, particularly for large positive pseudorapidity, where a one dimensional approximation is expected to hold.
2. There should exist a threshold in the coherent pion field intensity above which transparency should occur.
3. The suppression of cascading, if due to SIT, should be a periodic function of the intensity of the coherent meson field.
4. Early observations in cosmic ray physics and more convincing recent work at LBL[13] with emulsions has provided evidence for the existence, among the fragments arising in heavy ion collisions, of "anomalous" objects with a lifetime > 10^{-13} sec. and a mean free path several times smaller than that to be expected from normal nuclear interactions. It is conceivable that in heavy ion collisions the non-linear excitations produced could be metastable (soliton property) and could be carried off by an outgoing nuclear fragment. The interaction of such an excited fragment with ordinary nuclei might lead to a transverse instability[14] (growth) which could increase every time a collision takes place until eventually the excitation disintegrates. This would

lead to a decrease of the mean free path compared with what one would expect from "normal" events.

It would be very interesting to investigate whether there is indeed a connection between these experimental findings and the theoretical speculations put forward in this note.

We are indebted to E.M. Friedländer for very useful discussions.

REFERENCES

1. A recent experimental review on this subject can be found in T. Ferbel, University of Rochester, preprint C00.3065-236 (1979).
2. S.L. McCall and E.L. Hahn, Physical Review Letters 18, 908 (1967).
3. A recent theoretical review on this subject can be found in R.K. Bullough in "Interaction of Radiation with Condensed Matter". Volume 1, pg. 381, IAEA, ed. by R.K. Bullough and P.J. Caudrey.
4. J.C. Diels and E.L. Hahn, Physical Review A8, 1084 (1973).
5. R.A. Marth, D.A. Holmes and J.W. Eberley, Phys. Rev. A9 2733 (1974).
6. A.D. Vuzhva, Sov. Phys. Sol. State 20, 155 (1978).
7. O. Akimoto and K. Ikeda, J. Phys. A10, 425 (1977).
8. S.A. Moskalenko et al. Sov. Phys. Sol. State 19, 1271 (1977).
9. V.M. Agranovitch and V.N. Rupasov. Sov. Phys. Sol. State 18, 459 (1976).
10. J. Coll and H. Haken, Phys. Rev. A18, 2241, (1978).
11. G.N. Fowler and R.M. Weiner, Physics Letters 70B, 201 (1977), Phys. Rev. D17, 3118 (1978).
    G.N. Fowler, N. Stelte and R.M. Weiner, Nucl. Phys. A319, 349 (1979).
12. M. Deutschmann et al., CERN-preprint EP/PHYS. 78-1 (1978).
13. E.M. Friedländer, private communication (1979).
14. V.G. Makhankov, Phys. Rep. 35C, 2 (1978).

PROGRAM

ORBIS SCIENTIAE 1980

MONDAY, January 14, 1980

Opening Address and Welcome

SESSION I:	TOPICS IN ORBIS SCIENTIAE
Moderator:	Behram Kursunoglu, University of Miami
Dissertators:	P.A.M. Dirac, Florida State University "THE VARIATION OF G AND THE PROBLEM OF THE MOON"
	A.J. Meyer, The Chase Manhattan Bank "PRIMATONS, MAXIMUM ENERGY DENSITY QUANTA, POSSIBLE CONSTITUENTS OF THE YLEM"
	Maurice M. Shapiro, Naval Research Laboratory "THE DUMAND PROJECT AND HIGH ENERGY PHYSICS"
SESSION: II	DIVERS SPECTROSCOPIES
Moderator:	Sydney Meshkov, National Bureau of Standards
Dissertators:	M.A.B. Beg, Rockefeller University "DYNAMICAL HIGGS MECHANISM AND HYPERHADRON SPECTROSCOPY"
	Sydney Meshkov, National Bureau of Standards "GLUEBALLS: THEIR SPECTRA, PRODUCTION AND DECAY"
	Gabriel Karl, University of Guelph "QUARKS IN LIGHT BARYONS"
	O.W. Greenberg, University of Maryland "COLOR VAN DER WAALS FORCES?"
SESSION III:	A DAY OF EXPERIMENT
Moderator:	Nicholas P. Samios, Brookhaven National Laboratory
Dissertators:	Marvin Goldberg, Syracuse University "A. CLEO"
	J. Yoh, Columbia University "B. COLUMBIA-STONY BROOK"

	Samuel C.C. Ting, Massachusetts Institute of Technology "e^+e^- INTERACTIONS OBSERVED AT PETRA"
	Maurice Goldhaber, Brookhaven National Laboratory "THE QUESTION OF PROTON STABILITY, SUMMARY"
	Stephen L. Olsen, University of Rochester "A REVIEW OF CHARMED PARTICLE PRODUCTION IN HADRONIC COLLISIONS"
SESSION IV:	A DAY OF EXPERIMENT (continued from morning)
Moderator:	Nicholas P. Samios
Dissertators:	Neville W. Reay, Ohio State University "LIFETIME OF CHARMED PARTICLES"
	Richard Heinz, Indiana University "A SEARCH FOR NEW FLAVOR - BARE BEAUTY"
	Greg Donaldson, Stanford Linear Accelerator Center "DECAY BRANCHING RATIO OF CHARMED PARTICLES"
	Nicholas P. Samios, Brookhaven National Laboratory "WEAK AND ELECTROMAGNETIC PRODUCTION OF CHARMED BARYONS"
SESSION V:	CONFINEMENT
Moderator:	Fredrik Zachariasen, California Institute of Technology
Dissertators:	Fredrik Zachariasen, California Institute of Technology "INFRARED PROPERTIES OF THE GLUON PROPAGATOR: A PROGRESS REPORT"
	Roger Dashen, Institute for Advanced Studies "GAUGE THEORY COUPLING CONSTANT AT LARGE AND SMALL DISTANCES"
	George Zweig, California Institute of Technology "QUARK CHEMISTRY"
	Stanley J. Brodsky, Stanford Linear Accelerator Center "ASYMPTOTIC FREEDOM FOR EXCLUSIVE PROCESSES"
SESSION VI:	GRAND UNIFIED THEORIES
Moderator:	Pierre Ramond, California Institute of Technology
Dissertators:	William J. Marciano, Rockefeller University "THEORETICAL ASPECTS OF PROTON DECAY"

Richard Slansky, Los Alamos Scientific Laboratory
"FUN WITH E_6"

Leonard Susskind, Stanford University

Pierre Ramond, California Institute of Technology
"SO_{10} AS A VIABLE UNIFICATION GROUP"

SESSION VII: GAUGE AND OTHER FIELD THEORIES

Moderator: Sydney Meshkov, National Bureau of Standards

Dissertators: Paul M. Fishbane, University of Virginia
"MIGDALISM REVISITED: CALCULATING THE BOUND STATES OF QUANTUM CHROMODYNAMICS"

Pran Nath, Northeastern University
"THE U(1) PROBLEM AND ANOMALOUS WARD IDENTITIES"

Frederick M. Cooper, Los Alamos Scientific Laboratory
"RENORMALIZING THE STRONG-COUPLING EXPANSION FOR QUANTUM FIELD THEORY: PRESENT STATUS"

Geoffrey B. West, Los Alamos Scientific Laboratory
"ON THE n-p MASS DIFFERENCE IN QCD"

Boris Kayser, National Science Foundation
"TESTING STRONG AND WEAK GAUGE THEORIES IN QUARKONIUM DECAYS"

R.E. Marshak, University of Minnesota
"SELECTION RULES FOR BARYON NUMBER NONCONSERVATION IN GAUGE MODELS"

R.M. Weiner, Marburg University
"NONLINEAR EFFECTS IN NUCLEAR MATTER AND SELF-INDUCED TRANSPARENCY"

PARTICIPANTS

Hadi H. Aly
University of Colorado

Richard Arnowitt
Northeastern University

M.A.B. Beg
Rockefeller University

Carl M. Bender
Washington University

Stanley J. Brodsky
Stanford University

Arthur A. Broyles
University of Florida

Kevin Cahill
Harvard University

Richard H. Capps
Purdue University

Carl-Edwin Carlson
College of William and Mary

Roberto Casalbuoni
Istituto di Fisica Nucleare

William E. Caswell
University of Maryland

George Chapline
Lawrence Livermore Laboratory

Fred Cooper
Los Alamos Scientific Laboratory

Andreas P. Contogouris
McGill University

Bruce Cork
University of California
at Berkeley

John M. Cornwall
University of California
at Los Angeles

Richard H. Dalitz
Oxford University

Roger Dashen
Princeton University

P.A.M. Dirac
Florida State University

Greg Donaldson
Stanford Linear Accelerator Center

Hiroshi Enatsu
Ritsumeikan University, Kyoto

Albert Erwin
University of Wisconsin

Zyun F. Ezawa
City College of the City University
of New York

Gordon Feldman
Johns Hopkins University

Paul Fishbane
University of Virginia

Paul H. Frampton
Harvard University

J.A. Gaidos
Purdue University

Mary K. Gaillard
Lab. d'Annecy-le-Vieux de Phys. des Particles

David Garelick
Northeastern University

Marvin Goldberg
Syracuse University

Maurice Goldhaber
Brookhaven Laboratory

O.W. Greenberg
University of Maryland

Gerald S. Guralnik
Brown University

C.R. Hagen
University of Rochester

Vasken Hagopian
Florida State University

Arthur Halprin
University of Delaware

M.Y. Han
Duke University

Barry J. Harrington
University of California at Santa Barbara

Richard Haymaker
Louisiana State University

Richard Heinz
Indiana University

Peter Herczeg
Los Alamos Scientific Laboratory

George K. Kalbfleisch
University of Oklahoma

J.S. Kang
Rutgers University

Gabriel Karl
University of Guelph

Stewart Kasdan
Princeton University

Masaaki Kawaguchi
Kobe University, Japan

Boris Kayser
National Science Foundation

Nicola N. Khuri
Rockefeller University

Jihn E. Kim
University of Pennsylvania

Thaddeus F. Kycia
Brookhaven National Laboratory

Michael Longo
University of Michigan

F.E. Low
Massachusetts Institute of Technology

K.T. Mahanthappa
University of Colorado

Alfred K. Mann
University of Pennsylvania

William J. Marciano
Rockefeller University

R.E. Marshak
University of Minnesota

Mael A. Melvin
Temple University

Sydney Meshkov
National Bureau of Standards

A.J. Meyer
The Chase Manhattan Bank

R.N. Mohapatra
City College of New York

Thomas Nash
Fermi National Accelerator Laboratory

Pran Nath
CERN, Switzerland

Patrick J. O'Donnell
University of Toronto

Reinhard Oehme
University of Chicago

John R. O'Fallon
Argonne Universities Association

Stephan L. Olsen
University of Rochester

Jay Orear
Cornell University

Heinz R. Pagels
Rockefeller University

William F. Palmer
Ohio State University

Zohreh Parsa
New Jersey Institute of Technology

Trinang Pham
Centrede Physique Theorique Ecole Polytechnique, Paris

Enrico Poggio
Massachusetts Institute of Technology

P. Ramond
California Institute of Technology

Lazarus Ratner
Argonne National Laboratory

A.L. Read
Fermi Laboratory

Neville W. Reay
Ohio State University

Ronald Rockmore
Rutgers University

Nicholas P. Samios
Brookhaven National Laboratory

Howard J. Schnitzer
Brandeis University

Goran Senjanovic
University of Maryland

Maurice M. Shapiro
Naval Research Laboratory

Richard Slansky
Los Alamos Scientific Laboratory

A. Soni
University of California

J. Sucher
University of Maryland

George Sudarshan
University of Texas

Lawrence R. Sulak
University of Michigan

Leonard Susskind
Stanford University

Katsumi Tanaka
Ohio State University

Vigdor L. Teplitz
U.S. Arms Control and Disarmament

Samuel C. Ting
Massachusetts Institute of Technology

David G. Underwood
Argonne National Laboratory

Leon van Hove
European Organization for Nuclear Research, Geneva

Kameshwar C. Wali
Syracuse University

Richard Weiner
Marburg University, Germany

Geoffrey B. West
Los Alamos Scientific Laboratory

Gaurang B. Yodh
National Science Foundation

John Yoh
Columbia University

F. Zachariasen
California Institute of Technology

Bruno Zumino
CERN, Switzerland

George Zweig
California Institute of Technology

INDEX

Antiquarks 67ff
Atomic time 5ff
B-violating amplitudes 278
B-L violating decays 279
Baryons 61ff
excited 64
number 278
Baryon Number Nonconservation in Gauge Models, Selection Rules for 277-288
Base-Einstein correlations 297
Beam dump experiments 94-99
Beg, M.A.B. 23-31
Bender, Carl M. 211-238
Bessel function 177
Big-bang cosmology 125ff
Bjorken-Johnson-Low-limit 244
Black hole, Kerr 9ff
noncollapsing 10ff
rotating 13ff
time-reversed 14
Schwarzschild 18ff
Cartan subalgebra 151ff
Charmed Particles 89ff
mesons 89ff
Charmed Particle Production in Hadronic Collisions, A Review of 89-110
Coharmonium
spectroscopy 135,136
Color confinement 28
Color Van Der Waals Forces 67ff86
"Compton" radius 9ff
amplitude 241ff
Confinement 118ff
Cooper, Fred 211-238
Cottingham's formula 242ff
and the Bare Mass 252

Coulomb interaction 62ff;249
Dirac, P.A.M. 1-8
Dirac δ function 78
  octets 162
  quarks 162
Double well situation 225ff
Dynamical Higgs Mechanism and
  Hyperhadron Spectroscopy 23-31
Dynkin's representation theory 142ff
E_6 141ff
  Lie algebra 141ff
  Lie group 141ff
  mathematics of 142ff
  symmetry properties of 142ff
Earth-moon system 1ff
Einstein time 6ff
Electromagnetic radiation 9ff
Electroproduction experiments 136
Ephemeris time 6ff
Fishbane, Paul M. 173-188
Fowler, G.N. 289-298
Fun With E_6 141-164
Gauge
  bosons 123
  fields, ordinary 123
Gauge Theories in Quarkonium Decays,
  Testing Strong and Weak 257-276
Georgi-Glashow SU(5) model 121ff
Glueballs: Their Spectra, Production and Decay 43-60
Gluon vector dominance 93
Gluons 43ff
  propagator 111ff
Goldberger-Treiman constant 34
Goldhaber, Maurice 87-88
Goldstone bosons 29ff
Greenberg, O.W. 67-86
Green's functions 211ff
Guralnik, G.S. 211-238
Hadrons 68ff
  base 76ff
  exchange 69
  Fermi 76ff
  light 62
Higgs
  doublets 30
  field 166
  isodoublets 131
  mechanism 23ff
  multiplet 282

scalars 156;282
scheme 124
sector 40
Hilbert
space vector 143ff
Hyper-Scalar-Mesons 35
Hyper-Vector-Mesons 35
Hyperballs 45ff
Hyperbaryons 34ff
Hypercolor scenario 29
Hypergluons 45
Hyperhadron Spectroscopy 33ff
Hyperons 65
excited 63
Hyperpions 37ff
Hyperquarts 57ff
Hypersigmas 57
Infrared Properties of the Gluon Propagator: A Program Report 111
Interactions
coherent 289ff
Coulomb 62ff73
electromagnetic 141ff;277ff
flavor 151
meson-meson 291
mesonic 192
neutrino 92
proton 105
quark 62ff
quark hyperfine 65
strong 141ff;165;240ff,257ff;277ff
strong CP violating 194ff
Van der Waals analog 73
weak 141ff;165;240ff;257ff,277ff
Interquark interactions 61ff
Isaacson, work of 229ff
Isgur, Nathan 61-66
Jacobi identities 79
Jost function 178
Karl, Gabriel 61-66
Kayser, Boris 257-276
Kerr black hole 9ff
Kerr Newman model 10ff
Kink equation 221ff
Klein Gordon equation 291
Kronecker δ function 78
Lagrangian 194ff
QCD 194
Lattice field theory 212;224

Lattice Spacing, Extrapolating to zero 219ff
Lattice Strong Coupling Expansion 212ff
Lie group E_6 141ff
Majorana masses 166ff;283
Marciano, William J. 121-140
Marshak, R.E. 277-288
Mass M_S
the super-heavy 128ff
Mass predictions
super heavy 134
Meshkov, Sydney 43-60
Meson scattering,
quasi-elastic 84
Mesons 68ff
hadronic production of B 105
production 84
Meyer, A.J. 9-22
Migdalism Revisited: Calculating the Bound States of Quantum Chromodynamics 173-188
Minkowski space 217
Monte Carlo calculations 95;102
Moon, orbital motion of 2ff
problem of 1ff
Muon production, direct 101
N-p mass difference 239ff
N-p mass difference in QCD, On the 239-256
Nambu-Goldstone phenomenon 36
Nath, Pran 189-210
Neutrino beam dump experiment 94;99
Nonlinear Effects in Nuclear Matter and Self-Induced Transparency 289-298
Olsen, Stephen L. 89-110
Participants, List of 303-305
Pauli principle 67ff
Perturbation theory 72ff;115ff;174;228
chiral 192
Polarization, fixed beam 263
Preton-praton scattering 257ff
Primaton-Photon 9ff
Density 15ff
Primatons
Maximum Energy Density Quanta Possible Constituents of the Ylem 9-22
Program for Orbis Scientiae 1980 299-301
Proton Decay, Theoretical Aspects of 121-140
Proton Stability, The Question of 87-88
Quantum field theory 211ff
Quarks 26ff
anti- 67ff

base 73
Fermi 73
flavors 26
heavy 257ff
light 257ff
mass 27
mass splitting 239ff
Quarks in Ligh Baryons 61;66
QCD (Quantum Chromodynamics) 23ff;43ff;61ff;90
111;146;159;173ff;189ff;240ff
QED (Quantum Electrodynamics) 25ff;44;111;146ff;248;261ff
QFD (Quantum Flavordynamics) 23ff
Ramond, P. 165-172
Regge phenomenology 243
Renormalizing the Strong-Coupling Expansion
for Quantum Field Theory: Present Status 211-238
Roskies, Ralph 211-238
Scattering, proton-proton 257ff
Schrödinger equation 70ff
Sharp, David 211-238
Slansky, R. 141-164
SO_{10} as a Viable Unification Group 165-172
Spinology 111ff
SU_6 Breaking 65ff
Toponium 272
decay 259
U(1) Problem and Anomalous Ward Identities, The 189-210
Van der Waals forces 67ff
Variation of G and the Problems of the Moon, The 1-8
Violation
B- 278
B-L 279
CP-hard 125;167ff;189ff
parity 259
scaling 133
Ward Identities 192ff
Weinberg-Salam model 124
Weiner, R.M. 289-298
West, Geoffrey B. 239-256
Weyl reflection 146
White hole, nonexpanding 10ff
Yang-Mills theory 141ff
Yukawa couplings 166
Zachariasen, F. 111-120